Metals in Aquatic Systems:
A Review of Exposure, Bioaccumulation, and Toxicity Models

Other titles from the Society of Environmental Toxicology and Chemistry (SETAC):

*Bioavailability of Metals in Terrestrial Ecosystems:
Importance of Partitioning for Bioavailability to Invertebrates, Microbes, and Plants*
Allen, editor
2003

Contaminated Soils: From Soil–Chemical Interactions to Ecosystem Management
Lanno, editor
2003

Environmental Impacts of Pulp and Paper Waste Streams
Stuthridge, van den Heuvel, Marvin, Slade, Gifford, editors
2003

Porewater Toxicity Testing: Biological, Chemical, and Ecological Considerations
Carr and Nipper, editors
2003

Reevaluation of the State of the Science for Water-Quality Criteria Development
Reiley, Stubblefield, Adams, Di Toro, Erickson, Hodson, Keating Jr, editors
2003

*Bioavailability of Metals in Terrestrial Ecosystems:
Importance of Partitioning for Bioavailability to Invertebrates, Microbes, and Plants*
Allen, editor
2002

Community-Level Aquatic System Studies—Interpretation Criteria (CLASSIC)
Giddings, Brock, Heger, Heimbach, Maund, Norman, Ratte, Schäfers, Streloke, editors
2002

Interconnections between Human Health and Ecological Variability
Di Giulio and Benson, editors
2002

Silver in the Environment: Transport, Fate, and Effects
Andren and Bober, editors
2002

For information about SETAC publications, including SETAC's international journal *Environmental Toxicology and Chemistry*, contact the SETAC Administrative Office nearest you.

SETAC North America	SETAC Europe
1010 North 12th Avenue	Avenue de la Toison d'Or 67
Pensacola, Florida, USA 32501-3367	B-1060 Brussels, Belgium
T 850 469 1500	T 32 2 772 72 81
F 850 469 9778	F 32 2 770 53 83
E setac@setac.org	E setac@setaceu.org

www.setac.org

Environmental Quality Through Science®

Metals in Aquatic Systems:
A Review of Exposure, Bioaccumulation, and Toxicity Models

Paul R. Paquin
HydroQual, Inc.
Mahwah, New Jersey, USA

Kevin Farley
Manhattan College
Riverdale, New York, USA

Robert C. Santore
HydroQual, Inc.
Camillus, New York, USA

Christos D. Kavvadas
HydroQual, Inc.
Mahwah, New Jersey, USA

Kevin G. Mooney
General Electric Company
Pittsfield, Massachusetts, USA

Richard P. Winfield
U.S. Environmental Protection Agency
New York, New York, USA

Kuen-Bing Wu
HydroQual, Inc.
Mahwah, New Jersey, USA

Dominic M. Di Toro
University of Delaware
Newark, Delaware, USA

Coordinating Editor of SETAC Books

Andrew Green
International Lead Zinc Research Organization
Department of Environment and Health
Research Triangle Park, North Carolina, USA

Published by the Society of Environmental Toxicology and Chemistry (SETAC)

Cover design by Michael Kenney Graphic Arts and Design
Indexing by Celia McCoy

Library of Congress Cataloging-in-Publication Data

Metals in aquatic systems : a review of exposure, bioaccumulation, and toxicity models / Paul R. Paquin...[et al.].
   p. cm. -- (Metals and the environment series)
   Includes bibliographical references (p.).
   ISBN 1-880611-50-3
   1. Metals--Environmental aspects. 2. Aquatic ecology. I. Paquin, Paul R., 1950- II. SETAC (Society) III. Series.

QH545.M45M477 2003
577.6'2753--dc22   2003059242

Information in this book was obtained from individual experts and highly regarded sources. It is the publisher's intent to print accurate and reliable information, and numerous references are cited; however, the authors, editors, and publisher cannot be responsible for the validity of all information presented here or for the consequences of its use. Information contained herein does not necessarily reflect the policy or views of the Society of Environmental Toxicology and Chemistry (SETAC). Mention of commercial or noncommercial products and services does not imply endorsement or affiliation by the author or SETAC.

No part of this publication may be reproduced, stored in a retrieval system, or transmitted in any form or by any means, electronic, electrostatic, magnetic tape, mechanical, photocopying, recording, or otherwise, without permission in writing from the copyright holder.

All rights reserved. Authorization to photocopy items for internal or personal use, or for the personal or internal use of specific clients, may be granted by the Society of Environmental Toxicology and Chemistry (SETAC), provided that the appropriate fee is paid directly to the Copyright Clearance Center, Inc., 222 Rosewood Drive, Danvers, MA 01923 USA (Telephone 978 750 8400) or to SETAC. Before photocopying items for educational classroom use, please contact the Copyright Clearance Center (http://www.copyright.com) or the SETAC Office in North America (Telephone 850 469 1500, Fax 850 469 9778, E-mail setac@setac.org).

SETAC's consent does not extend to copying for general distribution, for promotion, for creating new works, or for resale. Specific permission must be obtained in writing from SETAC for such copying. Direct inquiries to the Society of Environmental Toxicology and Chemistry (SETAC), 1010 North 12th Avenue, Pensacola, FL 32501-3367, USA.

© 2003 Society of Environmental Toxicology and Chemistry (SETAC)
This publication was printed on recycled paper using soy ink.
SETAC Press is an imprint of the Society of Environmental Toxicology and Chemistry.
No claim is made to original U.S. Government works.

International Standard Book Number 1-880611-50-3
Printed in the United States of America
10 09 08 07 06 05 04 03   10 9 8 7 6 5 4 3 2 1

∞ The paper used in this publication meets the minimum requirements of the American National Standard for Information Sciences — Permanence of Paper for Printed Library Materials, ANSI Z39.48-1984.

Reference listing: Paquin PR, Farley K, Santore RC, Kavvadas CD, Mooney KG, Winfield RP, Wu K-B, Di Toro DM. 2003. Metals in aquatic systems: A review of exposure, bioaccumulation, and toxicity models. Pensacola, FL, USA: Society of Environmental Toxicology and Chemistry (SETAC). 168 p.

Metals and the Environment Series

This book is the second in the Metals and the Environment (MATE) Series, which originated during discussions of the Ecotoxicity Technical Advisory Panel (ETAP), a group sponsored by metals research organizations and committed to furthering research and understanding of metals in the environment.

The MATE Series is dedicated to the memory of Dr. Christopher M. Lee, who was a driving force for the creation of the ETAP and the creator of the concept for the MATE Series of publications. His leadership and vision on environmental issues will be missed, but his legacy provides a firm foundation for others to build upon.

SETAC Publications

Books published by the Society of Environmental Toxicology and Chemistry (SETAC) provide in-depth reviews and critical appraisals on scientific subjects relevant to understanding the impacts of chemicals and technology on the environment. The books explore topics reviewed and recommended by the Publications Advisory Council and approved by the SETAC North America Board of Directors and the SETAC World Council for their importance, timeliness, and contribution to multidisciplinary approaches to solving environmental problems. The diversity and breadth of subjects covered in the publications reflect the wide range of disciplines encompassed by environmental toxicology, environmental chemistry, hazard and risk assessment, and life-cycle assessment. SETAC books attempt to present the reader with authoritative coverage of the literature, as well as paradigms, methodologies, and controversies; research needs; and new developments specific to the featured topics. The books are generally peer reviewed for SETAC by acknowledged experts.

SETAC publications, which include Technical Issue Papers (TIPS), workshop summaries, newsletter (*SETAC Globe*), and journal (*Environmental Toxicology and Chemistry*), are useful to environmental scientists in research, research management, chemical manufacturing and regulation, risk assessment, life-cycle assessment, and education, as well as to students considering or preparing for careers in these areas. The publications provide information for keeping abreast of recent developments in familiar subject areas and for rapid introduction to principles and approaches in new subject areas.

SETAC recognizes and thanks the past SETAC editors:

    C.G. Ingersoll, Midwest Science Center
        U.S. Geological Survey, Columbia, Missouri

    T.W. LaPoint, Institute of Applied Sciences
        University of North Texas, Denton, Texas

    B.T. Walton, U.S. Environmental Protection Agency
        Research Triangle Park, North Carolina

    C.H. Ward, Department of Environmental Sciences and Engineering
        Rice University, Houston, Texas

Contents

List of Figures viii
List of Tables ix
Acknowledgments x
About the Editors xi
Executive Summary xv

Chapter 1 Introduction 1

Chapter 2 Overview of Aquatic Fate and Transport Models . 5
Overview of Model Frameworks Applied to Exposure and Risk
Assessments in Aquatic Settings 6
Scope of Literature Review 11

Chapter 3 Review of Aquatic Fate and Transport Models . 13
Literature Search Strategy 14
Review of Models 14
    Overview and description of models 14
    Comparison of fate and transport model features 35
Summary of Model Reviews and Guidance Documents 51

Chapter 4 Chemical Equilibrium Models 53
Historical Model Development 53
Comparison of Models 57

Chapter 5 Bioaccumulation and Toxicity Models 61
Introduction 61
Review of Bioaccumulation Models 62
    General equation for bioaccumulation 63
    Bioaccumulation models 65
    Application of bioaccumulation models 67
Review of Toxicity Models for Waterborne Metals 73
Modeling Metal Toxicity in Sediments 81
Integration of Bioaccumulation and Toxicity Models for Metals ... 87

Chapter 6 Model Selection and Future Model
Development Needs 91
Model Selection 91
    Hydrodynamic models 92
    Sediment transport models 92
    Fate and transport models 93
    Chemical equilibrium models 94
    Bioaccumulation and toxicity models 95

Future Model Development Needs 97
    Dynamic simulations and a probabilistic overlay 98
    Chemical equilibrium model 98
    Sediment chemistry model 99
    Toxicity model 100
    Bioaccumulation model 100
Concluding Remarks 101

Appendix: Partial List of Sources of Available Models ... 103
Abbreviations 107
References 113
Index 133

List of Figures

Figure ES-1	General fate and transport model framework	xvii
Figure ES-2	Biotic Ligand Model (BLM)-predicted LC50 versus observed LC50 for copper and silver	xxiv
Figure ES-3	Components of a modeling framework for aquatic ecological risk assessments (ERAs) for metals	xxv
Figure 2-1	General fate and transport model framework	7
Figure 2-2	Example modeling framework	8
Figure 3-1	Fluid transport regimes	38
Figure 3-2	Model dimensionality	40
Figure 3-3	Problem settling for zone of initial dilution and near field mixing zone analyses	41
Figure 3-4	A comparison of alternative particulate transport representations	43
Figure 3-5	Alternative partitioning approaches	46
Figure 3-6	A comparison of alternative representations of chemical partitioning and water–bed interactions	49
Figure 4-1	Schematic diagram of a chemical equilibrium model for metal speciation	53
Figure 5-1	Schematic of a 5-compartment generic food web model	66
Figure 5-2	Age-dependent striped bass model of the Hudson River estuary	67
Figure 5-3	Calculated and observed time history of total polychlorinated biphenyl (PCB) in the lake trout food chain for Lake Ontario	68
Figure 5-4	Comparison of observed biota–sediment accumulation factor (BSAF) and bioaccumulation factor (BAF) data of Oliver and Niimi to calibrated sculpin model	69
Figure 5-5	Schematic of pathways for metal uptake by bivalves	72
Figure 5-6	Comparison of model calibration and exposure data for *C. virginica* and *M. edulis*	72
Figure 5-7	Tissue accumulation of chromium in the rat after exposure to 100 ppm for 6 weeks	74
Figure 5-8	Schematic of 7-compartment pharmacokinetic model for cadmium in rainbow trout	74
Figure 5-9	Conceptual diagram of biotic ligand model	76
Figure 5-10	Biotic Ligand Model (BLM)-predicted LC50 versus observed LC50 for copper	77
Figure 5-11	BLM-predicted LC50 versus observed LC50 for silver	79
Figure 5-12	Equilibrium partitioning and water and sediment exposure	82
Figure 5-13	Dose–response curves based on porewater and carbon-normalized sediment concentrations	84
Figure 5-14	Comparison of dose–response curves for mortality as a function of bulk sediment metal concentration and simultaneously extracted metal/acid-volatile sulfide (SEM/AVS)	85
Figure 5-15	AVS and SEM model framework	86
Figure 5-16	Comparison of vertical profile data to SEM/AVS predicted with model	87
Figure 6-1	Components of a modeling framework for aquatic ecological risk assessments (ERAs) for metal	98

List of Tables

Table ES-1	List of fate and transport models	xx
Table ES-2	List of chemical equilibrium models	xxi
Table ES-3	List of bioaccumulation models	xxii
Table 3-1	List of fate and transport models	16
Table 3-2	Summary of features of analytical solution models	18
Table 3-3	Summary of features of steady-state numerical solution models	24
Table 3-4	Summary of features of time-variable numerical solution models	28
Table 3-5	Summary of features of fate and transport models	36
Table 4-1	Comparison of features of selected chemical equilibrium models for metal speciation	59
Table 5-1	Partial listing of hydrophobic organic chemical and metal applications for 5 bioaccumulation models	68

Acknowledgments

This report was produced in response to a request from the Organization of Economic Cooperation and Development (OECD), Task Force on Environmental Exposure Assessment, for input from the metals industry on exposure models for metals. As a result, the report was sponsored by the International Lead Zinc Research Organization (ILZRO), the International Copper Association (ICA), and the Nickel Producers Environmental Research Association (NiPERA) in coordination with the International Council on Metals and the Environment (ICME). Helpful comments were provided by the metal research associations, members of the OECD Task Force on Environmental Exposure Assessment, and two peer reviewers.

About the Authors

Paul Paquin, a Principal Engineer at HydroQual, Inc., Mahwah, New Jersey, USA, has nearly 30 years of experience in developing fate and transport models of organic chemicals and metals and in supervising many applications of these models in exposure assessments for natural waters. He has developed water and sediment quality modeling frameworks to assess the potential for effects of these constituents in aquatic settings. Mr. Paquin specializes in the development and application of procedures that consider contaminant bioavailability under field conditions, and he has participated in a number of scientific panels to consider related issues. He is an original co-developer of the biotic ligand model (BLM) of acute metal toxicity, an approach being adopted by the U.S. Environmental Protection Agency (USEPA) to set site-specific water quality criteria for metals, and he presented this approach to the USEPA Science Advisory Board. Mr. Paquin is currently working on a physiologically based model of metal bioaccumulation, with the goal of developing a framework that can be used to relate organ-specific metal accumulation levels that arise from waterborne and dietary exposure to effects.

Kevin Farley is an Associate Professor and Co-Chair of the Environmental Engineering Department, Manhattan College, Riverdale, New York, USA. He holds bachelor's and master's degrees from Manhattan College and a doctorate from MIT. Before returning to Manhattan College in 1995, he worked as an Assistant and Associate Professor at Clemson University and as a Senior Engineer for Tetra Tech, Inc. Dr. Farley's research focuses on water quality modeling and on the fate and bioaccumulation of toxic chemicals in aquatic systems, including chemical speciation and cycling of arsenic in surface waters and sediments, and modeling of chemical fate and bioaccumulation of polychlorinated biphenyls (PCBs), mercury, dioxins, and polyaromatic hydrocarbons (PAHs) in New York Harbor. He serves on the Model Evaluation Group for the Chesapeake Bay Program and has served on the National Research Council Committee on Remediation of PCB-Contaminated Sediments, the Lake Michigan Mass Balance Review Panel, the Science and Technical Review Panel for PCBs in the Hudson River, and as a water quality expert and consultant for the U.S. Department of Justice, the Confederated Salish and Kootenai Tribes, and the American Geological Institute. He is currently a consultant to HydroQual, Inc.

Robert Santore is an environmental scientist with more than 11 years of experience in the area of environmental and aquatic chemistry and chemical modeling. Mr. Santore developed the CHESS model to investigate the effects of atmospheric pollution on forest health and soil chemistry in the Adirondack Mountains of New York State. Recently, Mr. Santore has worked on several projects to evaluate the bioavailability and toxicity of heavy metals to aquatic organisms for HydroQual, Inc. in Camillus, New York, USA. This work has resulted in development of the BLM, which can explain and predict variations in the toxicity of several heavy metals to aquatic organisms.

Christos Kavvadas is an environmental engineer at HydroQual, Inc., Mahwah, New Jersey, USA, with more than 6 years of experience. He has been involved in the design, conduct, data analysis, and completion of both wastewater control studies and analyses of natural water systems that include a variety of water quality projects pertaining to both freshwater and marine systems. His duties include combined sewer-overflows field testing and data analyses; evaluation of measures for control of floatable materials; mathematical modeling of streams, estuaries, and reservoirs to evaluate the water quality impact of conventional and water quality pollutants; and data analyses for copper, silver, and zinc for evaluation of water quality criteria. In addition, Mr. Kavvadas is experienced in designing and organizing full-scale field experiments in wastewater treatment plants as well as bench-scale tests to investigate water quality characteristics for drinking and wastewater.

Kevin Mooney works for General Electric Company in Pittsfield, Massachusetts, USA, and has more than 13 years of experience studying environmental problems associated with the impact of conventional pollutants, organic chemicals, and metals in freshwater and estuarine systems through the application of mathematical and statistical analyses. Much of this work has included the development and application of mathematical models addressing hydrodynamics, sediment transport, chemical fate and transport, bioaccumulation, total maximum daily loads (TMDLs) and waste load allocation, and eutrophication or nutrient enrichment. Mr. Mooney's work has included permitting, stormwater management, environmental assessment, and natural resources investigations. More recently, he has focused on managing large-scale field and analytical programs incorporated in mathematical modeling studies and ecological and human health risk assessments to gain an understanding of the nature and extent of organic chemicals in natural aquatic and terrestrial systems.

Richard Winfield has been involved in environmental engineering studies since 1969, first in research followed by consulting, then industry, and currently the USEPA in New York, New York, USA. A broad range of physical settings were investigated in his water resource work, including those with hydraulic detention times ranging from a few hours to nearly a century. His primary focus has been water quality engineering analyses, including mathematical modeling of the physical, chemical, and biological components of river, estuary, coastal, and lake environments; field survey design; impact assessment; waste load allocations; new site selections; and siting and hydraulic design of outfall-diffuser systems. His involvement with the evaluation of fate and transport of toxic substances began in 1979. Since 1983, Mr. Winfield has developed engineering frameworks that address the fate, transport, bioaccumulation, and ecosystem effects of toxic substances (including PCBs, PCP, dioxin, and heavy metals). As a research consultant to Manhattan College, Mr. Winfield has worked with project teams that developed state-of-the-art water quality models for eutrophication and toxic substances, which were calibrated and verified using field data; the excellence of this research has been recognized by the International Association of Great Lakes Research and the American Society of Civil Engineering.

Kuen-Bing Wu, a Senior Project Manager at HydroQual, Inc., Mahwah, New Jersey, USA, has more than 15 years of experience in the fate and transport modeling of conventional pollutants, organic chemicals, and metals in natural water systems. His responsibilities have included model development and application, exposure and risk assessment, evaluation of remedial measures, and project management. He has conducted numerous estimated environmental concentration (EEC) analyses of agricultural pesticides undergoing Federal Insecticide, Fungicide, and Rodenticide Act (FIFRA) registration, having performed exposure analyses for more than 30 compounds. Mr. Wu has also analyzed waste load allocation, evaluated National Pollutant Discharge Eliminations System (NPDES) permit limits, and designed submerged outfall diffuser systems in both fresh and saline, tidal and non-tidal environments. He is currently working on projects involving the development of the BLM for a variety of metals, including copper, silver, cadmium, and nickel.

Dominic Di Toro, a principal consultant at HydroQual, Inc., and Distinguished Professor of Civil and Environmental Engineering at the University of Delaware, Newark, Delaware, USA, is an internationally recognized expert in the development and application of water quality models, for both conventional and nonconventional pollutants, including both organic chemicals and metals. Dr. Di Toro has served in a senior technical review capacity on USEPA's ongoing water quality criteria and sediment quality guideline development efforts, has published extensively in the scientific literature, and has participated as expert consultant and principal investigator on numerous studies for industry and governmental agencies. Dr. Di Toro has chaired or been an invited speaker at many scientific conferences. He provides technical oversight on several projects directed at the development of an exposure and effects assessment framework for copper, silver, and other metals, including a sediment submodel to predict fluxes of contaminants from sediments. In 1998, he received the Society of Environmental Toxicology and Chemistry (SETAC) Founders Award, the highest award made by SETAC for contributions to the field. Previously, Dr. Di Toro held the Donald J. O'Connor Chair of Environmental Engineering at Manhattan College.

Executive Summary

This report summarizes the results of a literature review and critical evaluation of models used in aquatic risk assessments for metals. An aquatic risk assessment includes both exposure assessment and effects assessment components. The purpose of the exposure assessment component is to characterize the predicted environmental concentrations (PECs) of metals in the environmental compartments of interest, such as the water-column, sediment, and biota of an aquatic system. The purpose of the effects assessment component is to summarize the data on the effects of the substance on the biota and to use these data to evaluate predicted no-effect concentrations (PNECs) in the environmental compartments of interest. The PNEC for a water-column compartment is frequently set equal to the water-quality criterion (WQC), should one exist. The potential for significant adverse impacts is assessed by comparing the PEC to the PNEC.

The specific objectives of this review were to
- identify and critique candidate fate and transport models for use in evaluating exposure levels of metals in surface waters and sediments;
- consider the utility of these models to evaluate metal bioavailability;
- identify and critique the candidate bioaccumulation and toxicity models for use with metals;
- evaluate the strengths and weaknesses of these models with respect to their use with metals and metal compounds;
- identify the most appropriate applications of existing models, given their current level of development; and
- identify weaknesses in the available modeling frameworks and recommend ways to improve the capabilities of current models.

The models considered in this review are broadly categorized as fate and transport models (Chapter 3), chemical equilibrium models (Chapter 4), and bioaccumulation and toxicity models (Chapter 5). Although not specifically included in this review, stand-alone hydrodynamic and sediment transport models are often employed in fate and transport analyses. As a result, several of these models will be briefly discussed as well.

While this review considers many of the better known modeling frameworks, as well as some that are not so well known, it was not possible to review all modeling frameworks that are available for use. Some of the models that have been included in this review were developed by the authors of this report or have been routinely used by them, and in this regard there may be an unintentional bias toward the selection of some of the models that have been included. To the degree this is so, it has arisen more out of an awareness of and familiarity with these models, rather than a conscious effort to exclude any particular model from the review. An exception to this is that the review does intentionally exclude models that were developed specifically for the purpose of evaluating the fate and transport, accumulation, and effects of metals such as mercury and selenium. Such models are relatively specialized due to the necessity to include processes such as methylation/demethylation reactions or, for metal species having a relatively high vapor pressure, volatilization (Masscheleyn and Patrick 1993; Zillioux et al. 1993). More generally, fate and transport models developed for specific metals that exist as organometallic forms and/or undergo changes in redox state (e.g., arsenic, chromium, mercury, and selenium) are not readily applicable to metals such as aluminum, cadmium, copper, nickel, lead, silver, and zinc. As a result, relatively specialized metal-specific fate and transport models developed for the former group of metals (e.g., the Mercury Cycling Model [MCM]; Hudson et al. 1994) are not considered in detail in this review. Rather, because many of the metals in the latter group are of direct interest to the sponsors of this review, our efforts have been directed at reviewing the types of models that are applicable to these metals.

The recommendations presented below have been developed based on this literature review. Recommendations are made concerning models to use in both fate and transport analyses of metals and in the evaluation of the bioaccumulation and toxicity of metals. Additionally, areas where future model development efforts should be directed to enhance the predictive capability of these models when applied to metals are indicated. A partial list of sources where models described in this review may be obtained is included in the Appendix.

Model Selection

Figure ES-1 illustrates the principal types of models that are commonly used in a fate and transport analysis and aquatic risk assessment and how these models are related in the context of an overall integrated modeling framework. While the development of an integrated approach to metal fate, bioaccumulation, and toxicity modeling was not a specific objective of this effort, it is useful to consider the information to be presented in this way because, as shown on Figure ES-1, there are clear links between the various steps in the overall modeling analysis. The development and application of an integrated approach to metal fate, bioaccumulation, and toxicity modeling presents a number of major challenges. Among the challenges are

1) the important interrelationships among the different processes must be included and properly represented (e.g., hydrodynamic results affect sediment transport results, which affect chemical fate and transport; chemical fate affects bioaccumulation and toxicity),
2) the theory is far from being completely understood, and
3) relevant models need to be linked in some manner, which implies that one model "upstream" of another may have to be modified to produce the full set of appropriate predictions.

It should be recognized at the outset that few if any situations involving metals in the environment have been modeled with complete success, and further, there are

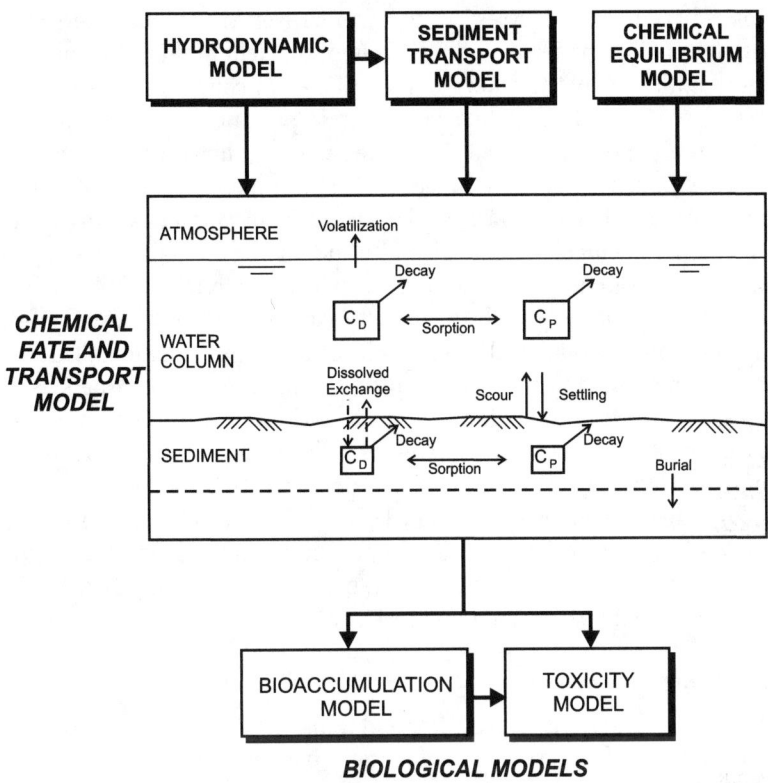

Figure ES-1 General fate and transport model framework

few if any situations where attempts have been made to model all of the aspects that will be discussed herein (i.e., simultaneous consideration of fate, bioaccumulation, and toxicity). At the same time, every step in the modeling process does not necessarily need to be performed each time that a model is applied. Often, data are used rather than model results as the input to any particular submodel. Hence, while an integrated modeling approach is of value, and it is useful to bear in mind the important physical, chemical, and biological relationships, it is also necessary that the various sub-models be suitable for use in a stand-alone mode as well.

This review does not consider in great detail the stand-alone hydrodynamic and sediment transport models that are sometimes used in advanced fate and transport analyses. However, recognizing that hydrodynamic and sediment transport models are likely to be used in some situations, we discuss these models briefly. This discussion is followed by recommendations pertaining to use of fate and transport models, chemical equilibrium models, and bioaccumulation and toxicity models.

Hydrodynamic models

Many fate and transport analyses incorporate relatively simple characterizations of water inputs and flow patterns, and this is often sufficient for modeling purposes. Of course, verification of the flows that are assigned by comparing computed conservative tracer results to field data is a very important step in the analysis, even when this simple approach is used. There are situations, however, where a more sophisticated modeling approach is needed. For example, use of a stand-alone hydrodynamic model may be warranted in an extremely complex setting where the model results have a high degree of visibility and important long-term ecological or economic implications. The Estuary, Coastal, Ocean Model (ECOM; Blumberg and Mellor 1987; Blumberg et al. 1993, 1999) is actually a family of state-of-the-art hydrodynamic models that would be suitable for use in this situation. ECOM employs a sophisticated turbulence closure algorithm to evaluate dispersive mixing and is capable of providing realistic time-variable currents even in complex systems. Other stand-alone hydrodynamic models that could be applied are the Environmental Fluid Dynamics Code (EFDC, Hamrick 1992) and the U.S. Army Corps of Engineers Chesapeake Bay model (CH3D, Johnson et al. 1993). Reliable predictions of currents may be critically important to the analysis because current-generated shear stresses control sediment transport processes, processes having a direct effect on the transport and ultimate fate of metals.

Sediment transport models

As is the case for fluid transport, the fate and transport analysis will often be based on a relatively simple representation of particle dynamics in the receiving water as well. The model inputs are typically set by calibration to water-column suspended solids data and long-term sedimentation rate data. There are situations, however, where a more mechanistically based sediment transport model such as ECOMSED (HydroQual 1998) or EFDC (TetraTech 2002b) may be warranted if reliable predictions are to be made beyond the range of model calibration conditions. For example, ECOMSED (HydroQual 1998) (described briefly in Chapter 3) has been developed to simulate both cohesive and noncohesive sediment transport and is capable of simulating armoring of the bed. Armoring is an important process to consider, because it can have a significant effect on predicted exposure levels. This is because the resuspension of bed sediment ceases when the surface of the sediment armors. This in turn has significant implications with respect to predicted exposure levels. The coupling of hydrodynamic and sediment transport models is important in regards to quantifying the impact of episodic events (floods, hurricanes, etc.) on contaminated sediment resuspension and redistribution.

Fate and transport models

Even when stand-alone hydrodynamic and sediment transport models are used in support of the exposure assessment, the fate and transport model is ultimately used to evaluate metal exposure levels. The specific fate and transport models reviewed herein are listed in Table ES-1. This table includes the model acronym, the model name, and a citation where additional information about each model can be obtained. There are advantages and disadvantages associated with each of the fate and transport models that are reviewed, and no single model is best for use in all cases. The following models, in order of increasing complexity, are suggested for use in various situations:

- The simplified procedures described in WQAM will serve as a useful starting point for making relatively simple calculations of exposure levels in screening-level analyses. This methodology is also suggested for use by individuals who are relatively inexperienced in working with fate and transport models.
- SLSA is an appropriate model to use in making a more mechanistically based evaluation of metal exposure levels in a one-dimensional (1-D) stream or river setting, or in a simple lake setting. MICHRIV is essentially the same model, but has additional flexibility with respect to representing a 1-D channel that has a variable cross-section.
- CTAP is suggested for use as a steady-state model when a multidimensional model of an irregularly shaped water body is needed. Typically, a model such as CTAP will be more time consuming to set up than a simpler model such as SLSA or MICHRIV.
- If a detailed time-variable analysis of a complex setting is needed, either WASP5 or WASTOX would be appropriate for use. These models call for a relatively high level of experience by the user, especially if used in conjunction with a state-of-the-art hydrodynamic or sediment transport model. As an alternative, DELFT3D also provides a refined modeling system that is applicable to complex settings.

When properly applied, each of these models should provide the analyst with a useful tool for performing an exposure assessment. An important limitation of most of the fate and transport models reviewed herein is that they do not provide a very refined approach to evaluating metal chemistry. Thus, it is recommended that they be used in conjunction with a chemical equilibrium model, discussed next, for purposes of evaluating speciation and complexation reactions. DELFT3D does include a chemistry submodel that is applicable to certain "standard applications."

Chemical equilibrium models

There are many chemical equilibrium models documented in the literature (Nordstrom et al. 1979; Bassett and Melchior 1990). Of the ones reviewed herein (Table ES-2), several have gained relatively widespread acceptance (most notably MINTEQA2 and MINEQL/MINEQL+; Westall et al. 1976; Schecher and McAvoy 1992) and are frequently used. Recently, the Windermere Humic Aqueous Model WHAM; Tipping 1994), has further advanced the capabilities of chemical equilibrium models in applications to natural water systems as a result of its standardized

Table ES-1 List of fate and transport models

Acronym	Model name	Reference
Analytical solution models		
WQAM	Water Quality Assessment Methodology	Mills et al. 1982a, 1982b, 1985
RIVRISK	River Risk	Grieb 1995; EPRI 1996
SLSA	Simplified Lake and Stream Analysis	Di Toro, O'Connor et al. 1981; HydroQual 1982a
GMIII	General Motors III	HydroQual 1982b; Ogden 1984
MICHRIV	Michigan River	Delos et al. 1984; Mills et al. 1985
DJOC	Donald J. O'Connor (1988), a triad of papers	O'Connor 1988a, 1988b, 1988c
QWASI	Quantitative Water Air Sediment Interaction	Mackay et al. 1983; Mackay 1991
USES	Uniform System for Evaluation of Substances (replaced by EUSES; EC 1996b)	RIVM et al. 1994, as described in Johnson and Luttik 1995
Steady-state numerical solution models		
CTAP	Chemical Transport and Analysis Program	HydroQual 1981, 1982a
PAWTOXIC	Pawtuxent Toxics	Wright 1987
SMPTOX3	Simplified Method Program-Variable-Complexity Stream Toxics	LTI 1992
	Version 2	Dilks et al. 1994, 1995
MEXAMS	Metals Exposure Analysis Modeling System Includes: EXAMS MINTEQ	Felmy, Brown et al. 1984 Burns et al. 1982 Felmy, Girvin, Jenne et al. 1984
Time-variable numerical solution models		
EXAMSII	Exposure Analysis Modeling System-II	Burns and Cline 1985 (Also, Burns et al. 1982; Burns 1990, 1997)
RIVEQLII	River Quality II	Chapman 1982
WASTOX	Water Quality Analysis Simulation of Toxics	Connolly and Winfield 1984
RCATOX (AESOP)	Row Column AESOP for Toxics (AESOP: Advanced Ecological Systems Operating Program)	HydroQual 2003
WASP5	Water Quality Analysis Simulation Program, Version 5	Ambrose et al. 1993
	Distributed with DYNHYD5, Dynamic Hydrodynamics 5	Ambrose et al. 1993
DELFT3D	Delft 3D Model	Delft Hydraulics 1998
HSPF	Hydrologic Simulation Program-FORTRAN	Donigian et al. 1984; Bicknell et al. 1993
CHNTRN	Channel Transport Model	Yeh 1981, 1982, as described in USEPA 1987
FETRA	Sediment/Radionuclide Transport Model	Onishi and Thompson 1984, as described in USEPA 1987
SERATRA	Sediment Contaminant (i.e., radionuclide) Transport	Onishi and Wise 1982a, 1982b
RECOVERY	RECOVERY	Boyer et al. 1994; Ruiz et al. 2000

Note: Several additional models that had not been identified for purposes of the initial detailed model review, but have since been included via brief description in Chapter 3, are as follows: 2 analytical solution models, EUSES, SYVAC, TRANSPEC/BIOTRANSPEC, META4, San Francisco Bay model (Chen 1996), and MIKE21.

Table ES-2 List of chemical equilibrium models

Model acronym	Model name	Reference
MINEQL	Minicomputer Equilibrium	Westall et al. 1976
MINTEQ	Minicomputer Water Equilibrium (Metal Speciation Equilibrium Model...)	Allison et al. 1991 or Felmy, Girvin, Jenne 1984
MINEQL+	Minicomputer Water Equilibrium Model, +	Schecher and McAvoy 1992
WHAM	Windermere Humic Aqueous Model	Tipping 1994
CHESS	Chemical Equilibrium in Soils and Solutions	Santore and Driscoll 1995
NICA	Non-Ideal Competitive Adsorption	Bennedetti et al. 1995 or Kinniburgh et al. 1996

calibration of metal–organic matter interactions to a variety of published data. The Chemical Equilibrium in Soils and Solutions (CHESS) Model is a more recently developed chemical equilibrium model that is also described in this review (Santore and Driscoll 1995). The Biotic Ligand Model (BLM), used to predict metal toxicity and described subsequently, is based on CHESS, modified to incorporate the metal–organic matter interaction approach employed in WHAM. Based on a feature-by-feature comparison of these models, many similarities and some important differences are evident. Although any of these models could be used effectively in an environmental exposure and risk assessment for metals, the determination of the "best" model to use must be made on a case-by-case basis. The choice of model will depend on what features are most important for a given application and for a given user. Both MINEQL+ and MINTEQA2 are powerful, general-purpose equilibrium models and are supplied with actively maintained databases of thermodynamic constants. Of these two, MINEQL+ has undergone recent updates, including an easy to use menu-driven interface and expansion of its thermodynamic database. On the other hand, WHAM specializes in interactions between metals and natural organic matter in an aqueous environment. WHAM is less comprehensive than MINTEQA2 or MINEQL+ and does not include solubility or redox transformations. However, WHAM may be more suitable for applications where interactions between metals and dissolved organic matter are important. The Non-Ideal Competitive Adsorption model (NICA: Koopal et al. 1994; Benedetti et al. 1995; Kinniburgh et al. 1996; Temminghoff et al. 1997; de Rooij et al. 1999) may serve as an alternative in this regard. NICA, discussed briefly herein, was developed by the Delft Hydraulics Laboratory and has a structure that is comparable to WHAM.

Bioaccumulation and toxicity models

The key bioaccumulation models reviewed herein are listed in Table ES-3. These models have been developed over the previous 25 years to describe the processes of contaminant uptake, depuration, and transformation in aquatic organisms and contaminant transfers through aquatic food webs. Of these models, the Thomann

Table ES-3 List of bioaccumulation models

Model acronym	Model name	Reference
	Thomann	Thomann 1977, 1978, Thomann et al. 1984, 1992a, 1992b, 1995, 1997; Connolly and Thomann not dated
	Gobas	Gobas 1993; Gobas et al. 1995
AQUATOX	Aquatic Toxicity	Park 1998; USEPA 2000
BASS	Bioaccumulation and Aquatic System Simulator	Barber 1999
FGETS	Food and Gill Exchange of Toxic Substances	Barber et al. 1991; Suarez and Barber 1994

and Gobas bioaccumulation models have been widely applied and are considered to be most generally accepted by the scientific community. These two models yielded similar results when compared in a steady-state analysis of polychlorinated biphenyls (PCBs) in a Great Lakes food chain (Burkhard 1998). Compared to hydrophobic organic chemicals (HOCs), examples of bioaccumulation model applications to metals are much more limited in extent, and these applications have focused on the bioaccumulation dynamics of individual aquatic species rather than aquatic food webs (e.g., Reinfelder et al. 1997). Significant differences in bioaccumulation at different trophic levels and by individual aquatic species of the same trophic level have been observed and are not well understood at this time. As a result, further developmental work is needed in this area. Even so, the current version of either the Thomann model or the Gobas model will provide the analyst with useful tools for assessing metal bioaccumulation. The Thomann model has been more widely applied to metals to date.

With regard to the assessment of water-column and sediment toxicity, the importance of considering bioavailability has been demonstrated both for nonionic organic chemicals (Di Toro et al. 1991; USEPA 1989, 1993; Ankley, Berry et al. 1996) and for metals (Di Toro et al. 1992; Ankley et al. 1993, 1994; Ankley, Di Toro 1996; USEPA 1994b; Kramer et al. 1997; Renner 1997). Metal bioavailability and toxicity in the water-column have long been recognized as being a function of water chemistry (e.g., Pagenkopf et al. 1974; Sunda and Guillard 1976; Sunda and Hansen 1979). The formation of inorganic and organic metal complexes and sorption of metals to particle surfaces have been shown to reduce toxicity. As a result, the relationship of metal toxicity to total or dissolved concentrations can be highly variable, depending on ambient water chemistry. As explained in the context of the Free Ion Activity Model (FIAM), in comparison to these measurements, the free metal ion concentration provides a much better indication of bioavailability and toxicity (Sunda and Guillard 1976; Morel 1983). Allen and Hansen (1996) showed how metal speciation is expected to affect toxicity and, using chemical equilibrium calculations, predicted the range of effects on copper toxicity resulting from the variation of site-specific water quality characteristics for a number of water bodies.

The BLM was developed to incorporate metal speciation and the protective effects of competing cations into predictions of water-column metal bioavailability and toxicity (Di Toro et al. 1997, 2001; USEPA 1999a). The BLM is based on a conceptual model that is similar to the gill site interaction model proposed by Pagenkopf (1983) and to the FIAM as described by Morel (1983). Although models capable of predicting metal bioaccumulation on the gill in short-term exposures have been previously developed (Playle et al. 1993a, 1993b), the BLM is the first to employ this approach in conjunction with a scheme to predict a toxic effect level. While still in the developmental stages, the BLM holds great promise as a way to predict the effects of site-specific water quality on toxicity. It has been used to predict the acute toxicity of both copper (fathead minnows and *Daphnia pulex*) and silver (fathead minnows, rainbow trout, and *Daphnia magna*) to within a factor of 2 over a range of water quality conditions (Figure ES-2). The BLM has also been applied to the analysis of copper toxicity data for the blue mussel, *Mytilus edulis*, a saltwater species. A number of applications to other metals, including cadmium, nickel, and zinc, and to other organisms have been reported (see Paquin, Gorsuch et al. 2002 for a review.) However, further testing and model development are needed before it can be routinely applied to other organisms and other metals, in both freshwater and marine settings. In the interim, use of dissolved metal concentrations and metal speciation calculations can also be used to obtain an indication of available metal.

The equilibrium partitioning (EqP) approach for assessing sediment toxicity, originally developed for nonionic organic chemicals, was subsequently extended to metals, including copper, cadmium, nickel, lead, zinc, and silver (Di Toro et al. 1992; Ankley, Di Toro et al. 1996; Berry et al. 1996, 1998; USEPA 1999a). Specifically, the approach is based on the assumption that the activity of these metals is controlled by the amount of acid-volatile sulfide (AVS) that is present. This is because AVS is expected to react with the labile fraction of these co-existing metals to form an insoluble metal-sulfide complex, one that is relatively nonbioavailable. Thus, when the measured AVS exceeds the concentration of the simultaneously extracted metal (SEM), the porewater levels of free metal should approach very low concentrations and metal toxicity is not expected (Di Toro et al. 1990, 1992; Ankley, Berry et al. 1996). When the ratio exceeds unity, this condition no longer applies and the magnitude of elevated levels of metals in the pore water becomes dependent upon the magnitude of the excess SEM (SEM–AVS), the concentrations of other ligands that are present in the pore water, and the degree of binding of the metals to particulate phases in the sediment. Use of carbon-normalized excess SEM is a recent refinement to this approach that begins to consider these other phases. A modeling framework for the evaluation of AVS and SEM in sediments, one that can be used to assess metal bioavailability and toxicity in sediments, is also described herein.

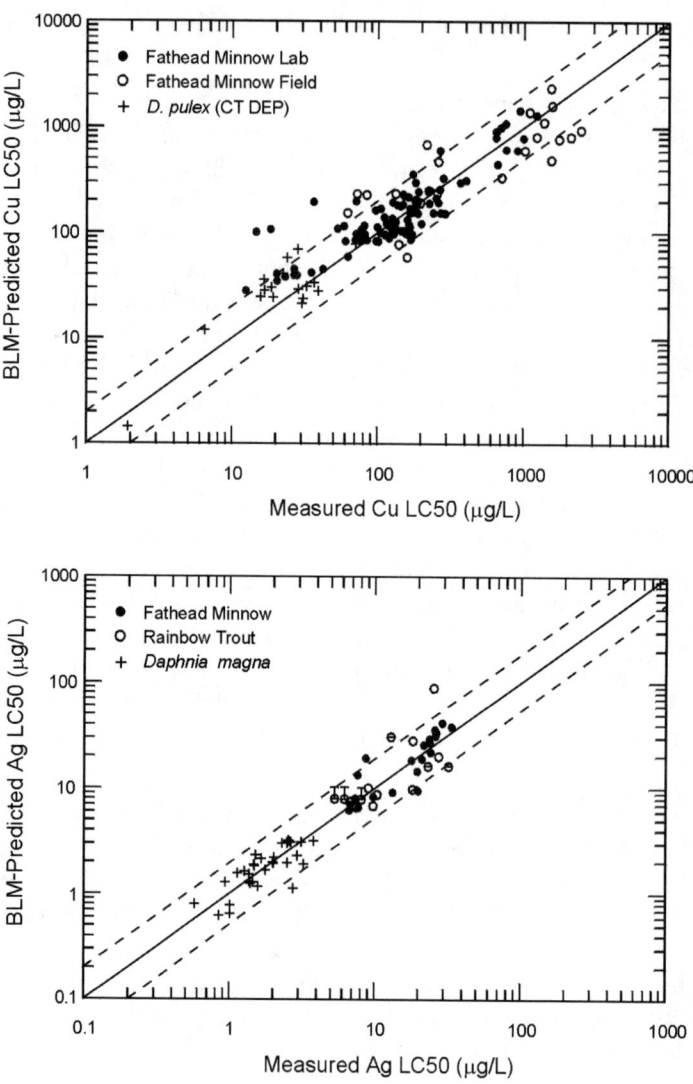

Figure ES-2 BLM-predicted LC50 versus observed LC50 for copper and silver

Future Model Development Needs

The models reviewed herein will serve as useful tools to the analyst evaluating the fate and effects of metals in aquatic environments. However, further model development efforts are warranted in a number of areas. Many of these same areas were also recognized as such at the 1997 Pellston Workshop (Bergman and Dorward-King 1997). Although significant advances have been made in the short period since the 1997 Pellston workshop was held, much remains to be accomplished.

Figure ES-3 illustrates some of the important model components that should be included in a comprehensive aquatic exposure and risk assessment framework for metals. The fate and transport model itself is represented by the water-column and sediment compartments shown at the center of the diagram. As shown to the left of Figure ES-3, a comprehensive fate and transport model framework will require the capability to perform dynamic simulations using temporally varying inputs. The time-variable output that results will be suitable for use in a probabilistic analysis of exposure and effects.

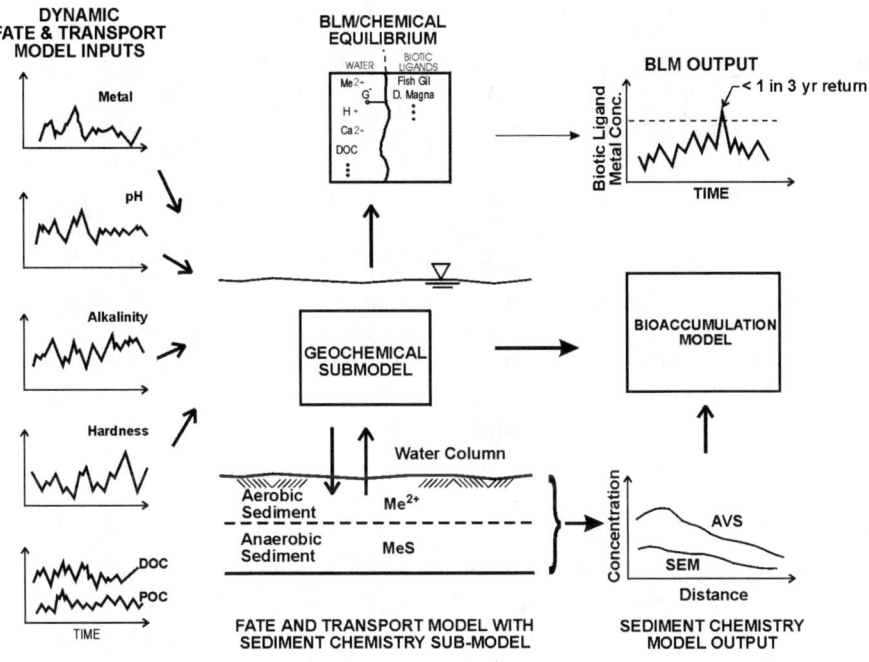

Figure ES-3 Components of a modeling framework for aquatic ERAs for metals

A comprehensive fate and transport model should also simulate the basic parameters needed by an integrated geochemical submodel (also shown at the center of the diagram). The geochemical model should provide an improved representation of metal speciation and partitioning in comparison to the relatively simple methods incorporated in most models currently in use today. A sediment chemistry submodel (middle and lower right diagrams), capable of simulating the interactions of metals with sulfide and other binding phases, is also needed to provide a proper basis for evaluating interaction between the water-column and sediment and to assess metal bioavailability and toxicity in sediments.

The improved representation of chemical speciation that is called for above is consistent with the level of detail provided by the BLM (upper middle and upper right diagrams of Figure ES-3). The BLM is used to assess bioavailability and toxicity as a function of site-specific and time-varying water quality. Finally, a multi-compartment pharmacokinetic-based bioaccumulation model (middle right diagram), one that reflects metal essentiality, sequestration, and trophic level transfer, is needed to complete a comprehensive exposure and risk assessment analysis. Each of these components is discussed briefly in the remainder of this section.

Dynamic simulations with a probabilistic overlay

When steady-state analyses are completed, regulatory agencies often require that critical low flow conditions and peak mass discharge rates be assumed. The concurrent application of multiple low probability events typically results in unnecessarily conservative results. Although in some instances this type of screening-level analysis is sufficient to rule out the likelihood of a problem occurring, in other instances it can lead to an inefficient utilization of resources. Thus the capability to conduct dynamic simulations, including Monte Carlo generation of time-variable inputs (shown on the left of Figure ES-3), will be needed if a sufficiently detailed characterization of exposure levels and effects is to be made. This capability is particularly important in a regulatory setting where a probabilistic analysis of the frequency of exceedance of a WQC is required to assess compliance. As discussed in Chapter 3 (Probabilistic overlays), a number of models having these capabilities are currently available for use.

Chemical equilibrium model

A chemical equilibrium model suitable for characterizing the water-column chemistry in a risk assessment for metals (included in the middle diagram of Figure ES-3) is also needed. It should include inorganic speciation, complexation with dissolved organic matter, adsorption on suspended particles, and solid-phase solubility constraints. Although some of the available models are adequate, an updated procedure for simulating sorption of metals to particles is still needed. Integration of a chemical equilibrium subroutine into a fate and transport model will ultimately enhance the predictive capability of the model as well. When warranted, a computationally efficient alternative to an integrated chemical equilibrium submodel might be to link the fate and transport model to a stand-alone geochemical model, without completing the detailed chemical equilibrium computations at every time step. This latter approach would require a demonstration that it yields results that are consistent with the integrated modeling approach.

Situations may also exist where nonequilibrium conditions prevail, such as in the immediate vicinity of a discharge. When this occurs, the available chemical equilibrium models may not be suitable for use. As a result, a long-term goal of future research and model development efforts should be to improve the predictive capabilities of both geochemical and fate and transport models under nonequilibrium conditions.

Sediment chemistry model

The basis for the EqP methodology was described previously and is discussed in further detail in Chapter 5 (Modeling Metal Toxicity in Sediments). Briefly, the EqP methodology as applied to metals includes consideration of SEM and how it interacts with AVS and other binding phases (e.g., dissolved organic carbon [DOC] and particulate organic carbon [POC]) as a way to evaluate metal bioavailability. A model capable of simulating the temporal and vertical variation of AVS and SEM in sediments (lower middle and lower right diagrams of Figure ES-3) is therefore needed for risk assessment purposes.

The oxidation of metal sulfides in surficial sediments is believed to be an important factor that affects the transfer of some dissolved metals between the water-column and sediment. An improved understanding of this process and the kinetics of metal sulfide oxidation in sediments is needed. Similarly, oxidation rate information is needed to evaluate the impacts of short-term sediment resuspension events that occur during high flow periods. Development of a sediment submodel capable of simulating these processes will improve both the short-term and the long-term predictive capability of fate and transport models for metals. A sediment model having both aerobic and anaerobic sub-layers has been developed and applied to cadmium (Di Toro, Mahony, Hansen, Berry 1996; Di Toro 2001) and is currently being developed for copper, silver, and other metals.

Prediction of toxic effects: The BLM of acute toxicity

An acute toxicity model for metals should be capable of accounting for the wide range in toxic effect levels that result from changes in water chemistry (including changes in pH, concentrations of dissolved organic matter (DOM) and total suspended solids (TSS), concentrations of hardness cations, and availability of inorganic ligands such as carbonate ions). An example of such a model is the BLM (Meyer et al. 1999; Paquin et al. 1999; USEPA 1999a; Di Toro et al. 2001; Santore et al. 2001). The BLM was originally developed using copper and silver toxicity data for freshwater fish and invertebrates. As described in Chapter 5, it has subsequently been extended to other metals and other organisms by numerous investigators (see Paquin, Gorsuch et al. 2002 for a review.) The BLM should be further validated using additional independent toxicity data sets for these same organisms, and the model parameter values refined as needed. It should also be extended in applicability to additional freshwater and saltwater organisms, for a variety of metals. Further studies should be conducted to assess the applicability of this general approach to the evaluation of chronic toxicity.

Bioaccumulation model

Examples of modeling studies of metal bioaccumulation are limited in number and have largely focused on analysis of laboratory data for specific aquatic organisms. Further development of bioaccumulation models for metals beyond the use of empirically derived bioconcentration factors is clearly needed (middle right panel of Figure ES-3). Particular attention should be given to multi-compartment pharmacokinetic models. This will lead to an improved understanding of how metals are sequestered in specific organs and how this sequestering affects depuration kinetics, toxicological effects on specific target organs, and bioavailability of metals to higher trophic-level organisms.

Significant differences in metal bioaccumulation are expected for different trophic levels and among species of the same trophic level. Reasons for this behavior are not well understood at this time, and further research is needed. Similarly, an improved understanding of the processes that control the active uptake and elimination of essential metals and how to best represent these processes in bioaccumulation models is also needed. Further advances in understanding in regard to intracellular speciation of metals, and how this affects metal detoxification and trophic transfer potential, should also be of use in future bioaccumulation model developments. Incorporation of refinements in these areas in existing bioaccumulation models will ultimately lead to an improved ability to model the bioaccumulation of metals both across species at a given trophic level and across trophic levels in more complex aquatic food webs. Finally, because metals often coexist with other metals and HOCs in the environment, a long-term goal in bioaccumulation modeling should be the assessment of synergistic effects among contaminants.

Concluding Remarks

Significant advances have been made since the time of the 1997 Pellston Workshop on WQC for metals, when needs were identified with regard to the development of methods for modeling metal transport, fate, accumulation, and effects. However, further developments are still needed. This is demonstrated by the fact that many of the areas described above were also topics of discussion at a May 2003 Pellston Workshop (Pensacola, Florida, USA) on the hazard identification approach for metals. It is expected that steady progress will continue to be made as the scientific community responds to the expressed needs of both government and industry.

CHAPTER 1

Introduction

This report summarizes the results of a literature review and a critical evaluation of models used in aquatic risk assessments for metals. The exposure assessment component of the risk assessment is completed to obtain a characterization of the concentrations of metals in the water column, sediment, and biota of aquatic environments. When these exposure concentrations are used in conjunction with concentrations known to result in adverse effects to the biota, the potential for significant adverse impacts may be assessed.

The procedures to follow in completing an aquatic risk assessment for organic chemicals and metals are described in any of a number of publications (e.g., Suter 1993; Cowan et al. 1995; EC 1996a). Although the details of the procedures may vary, the overall approach is generally consistent with the approach described in the European Commission's (EC) Technical Guidance Document (TGD) (EC 1996a). Pursuant to the TGD, calculated or measured concentrations of an organic chemical or a metal can be used to define the predicted environmental concentration (PEC), the exposure concentration of interest. Because exposure concentrations vary over time and space, the PEC is not necessarily uniquely defined. A range of PEC values may arise from limitations of the model used to predict these concentrations or uncertainty in modeling assumptions, as well as from characteristically observed variations in monitoring data.

The PEC range is compared to the predicted no-effect concentration (PNEC) to assess the potential for adverse ecological effects. The PNEC is regarded as a concentration below which an unacceptable effect will most likely not occur (EC 1996a). In accordance with TGD procedures, the PNEC is calculated by dividing the lowest short-term LC50 (lethal concentration for 50% of test organisms) or EC50 (effect concentration for 50% of test organisms) or long-term no-observed-effect concentration (NOEC) by an appropriate assessment factor. The magnitude of this factor depends on the level of confidence in the ecological test results used to define the PNEC. A lower factor is used if the PNEC is based on NOECs from 3 or more long-term studies (assessment factor = 10) than if it is based on one or 2 NOECs (factor = 50 to 100) or on a short-term LC50 or EC50 (factor = 50 to 1000). It is not the intent of applying these assessment factors to achieve a 100% level of protection but rather to protect most of the species in the aquatic ecosystem. The risk characterization is evaluated in terms of the ratio of the PEC to PNEC. If the ratio is greater than 1, the substance is of concern and further action has to be taken, including possible refinement of the PEC or PNEC. The magnitude of the PEC to PNEC ratio and the production level of the substance are considered in deciding how to proceed. Other factors to consider are the bioaccumulation potential of the substance, the shape of the toxicity–time curve, and data on structurally analogous substances.

In the Netherlands, statistical methods are applied to define a "level of no concern," a level that is consistent with the Dutch concept of a "maximum permissible level" that will protect at least 95% of the species (Johnson and Luttik 1995). If the PEC-to-PNEC ratio is greater than 1, the risk is considered high, between 0.01 and 1 it is considered medium, and if the ratio is less than 0.01, it is considered low.

The procedure that is often followed in the U.S. is to compare the predicted exposure concentrations to the PNEC, where the PNEC is typically set equal to the water-quality criterion (WQC) for the constituent of interest. The magnitude of the WQC, which is often available for metals, is also based on a 95-percentile level of protection (Stephan et al. 1985). The national WQC also specifies an allowable frequency and duration of exceedance of the WQC. It is because of these requirements that it is necessary to perform time-variable simulations, the output being amenable to analysis and evaluation of the magnitude, duration, and frequency of exceedance of the WQC. The results can also be used to evaluate the dissolved metal concentration, the form that is frequently used to evaluate the bioavailable fraction of the metal, and to develop permit limits. Residue-based WQC are developed for chemicals that have significant bioaccumulation potential.

Bioassay-based procedures have also been established to evaluate site-specific criteria for metals in order to consider the effect of site-specific water quality on bioavailability and toxicity (USEPA 1994a). The Biotic Ligand Model (BLM), discussed in Chapter 5, is being developed to provide a computational alternative to these procedures for setting site-specific WQC for metals. It is envisioned that this model will add a new perspective to the PEC-to-PNEC analysis for metals in general because it provides a way to predict toxic effect levels that reflect the chemistry of the receiving water. A dissolved concentration that results in a significant adverse ecological effect in one water body may not cause a significant effect in another, because the bioavailability of the metal may differ in the 2 water bodies. As such, consideration of the effect of site-specific water quality on toxicity may lead to the need to evaluate a site-specific PNEC that reflects conditions at the site of interest.

Numerous models are available for evaluating metal exposure levels for use in risk assessments. Of these, many were originally developed to predict the fate and transport of organic compounds and have subsequently been applied to metals. When this is done, the resulting characterization of metal speciation is inevitably oversimplified. This is a significant weakness in these types of models, because it is well known that metal speciation is a key factor in assessing the bioavailability and toxicity of metals (e.g., Sunda and Guillard 1976; Pagenkopf 1983; Campbell 1995; Allen and Hansen 1996). Chemical equilibrium models represent another class of models that were developed to provide a detailed chemical speciation characterization, but are of limited utility with respect to simulating metal fate and transport. In the more refined analyses, both of these types of models are used together. Resulting exposure levels form the basis of predictions of bioaccumulation and toxicity, and models are available to facilitate these evaluations as well.

The specific objectives of the literature review were to
- identify and critique candidate fate and transport models for use in evaluating exposure levels of metals in surface waters and sediments,
- consider the utility of these models to evaluate metal bioavailability,
- identify and critique the candidate bioaccumulation and toxicity models for use with metals,
- evaluate the strengths and weaknesses of these models with respect to their use with metals and metal compounds,
- identify the most appropriate applications of existing models, given their current level of development, and
- identify weaknesses in the available modeling frameworks and recommend ways to improve the current model capabilities.

Note that models developed to represent the unique characteristics of some metals, such as mercury, selenium, arsenic, and chromium, where methylation and demethylation reactions and/or changes in redox state may occur, are relatively specialized and hence have been intentionally excluded from consideration herein. Even so, many of the concepts that are discussed here are relevant with regard to consideration of these metals as well.

The results of the literature review as it pertains to these objectives are discussed in the remainder of this report. The report is structured as follows. As a prelude to the literature review, Chapter 2 provides an overview of the essential components of a fate and transport modeling analysis. Important features that should be considered in evaluating the models to be reviewed are discussed. The results of the literature review are summarized in Chapters 3, 4, and 5. The fate and transport models are reviewed in Chapter 3. Because many of these models were not explicitly developed for metals, the chemistry component, including partitioning to suspended particulate matter and metal speciation, is often represented simplistically. In this chapter, the methods used by various models to evaluate partitioning between dissolved and particulate phases and chemical reactions will be compared, and the ways that some models have been linked to more refined chemical equilibrium models discussed. The relatively refined approach to predicting metal partitioning, speciation, and complexation in the context of chemical equilibrium models are then discussed in detail in Chapter 4. These chemistry models are important because, though not designed as fate and transport models, they are sometimes used in conjunction with fate and transport models when applied by an experienced analyst. The model review concludes with a discussion of the applicability of various models for use in evaluating the bioaccumulation and toxicity of metals (Chapter 5).

Chapter 6 of this report consists of 2 parts. First, recommendations are made regarding the selection of models for use from the existing suite of available models. It is noted at the outset that no single model is appropriate for use in all cases. A number of individual models may actually be appropriate for use in some instances, depending on the problem setting and the objectives of the analysis. Alternatively, in some cases it may be necessary to use several models in conjunction with each other. At the same time, the state of development of fate and transport models in general, and models for metals in particular, is evolving. Areas where future development efforts should be directed are discussed in the remainder of Chapter 6. A partial list of sources for obtaining many of the models that are discussed in this review is presented in the Appendix. The report concludes with a complete list of cited references.

CHAPTER 2

Overview of Aquatic Fate and Transport Models

Fate and transport models developed for use in exposure and risk assessments may vary from very simple 1-dimensional (1-D) continuous analytical solution models to very complex 3-dimensional (3-D) numerical solution models. Although in their simplest form the model computations can be performed with a hand calculator, a more refined and detailed analysis of a complex setting may require application of an advanced model and supercomputer capabilities. Regardless of the situation, however, each of these models generally incorporates, with varying levels of sophistication and detail, certain basic components that control metal fate and transport. This chapter presents a brief description of the principal components of the fate and transport, chemical equilibrium, and bioaccumulation and toxicity models to be reviewed. Some of the important factors to be considered in deciding on the specific model to use for a particular application will also be discussed.

This initial, general discussion of models is intended to introduce the reader to some of the important concepts that will be referred to in subsequent sections of this report. A number of excellent review articles and textbooks that include more detailed descriptions of the conceptual basis and formulation of fate and transport models than is presented herein are available for the interested reader (Burns 1983; Medine and McCutcheon 1987; Schnoor et al. 1987; Thomann and Mueller 1987; Chapra 1997). Schnoor (1996), in a textbook on fate and transport of pollutants in water, air, and soil, devotes a chapter to modeling of trace metals. The interested reader is also encouraged to refer to papers by O'Connor (1988a, 1988b, 1988c) on models of sorptive substances in freshwater systems for an excellent introduction to this subject. O'Connor was a pioneer in the development of water quality modeling techniques, and this series of papers provide a concise description of representative processes and formulations employed in many of the fate and transport models reviewed herein.

Most fate and transport models were developed for application to neutral organic chemicals but have subsequently been applied to metals. Because the focus of this review is the applicability of these models to metals, the convention will be to refer to the constituent being modeled with the more generic terms "chemical" or "substance." More explicit terms such as "organic compound" or "metal" will generally be reserved for use when the discussion specifically applies to these types of constituents.

Overview of Model Frameworks Applied to Exposure and Risk Assessments in Aquatic Settings

Figure 2-1 illustrates the primary components and interrelationships of the models typically used to conduct exposure and risk assessments in aquatic settings. A schematic diagram showing some of the processes included in a representative fate and transport model is shown at the center of this figure. Although the details will vary, most models incorporate algorithms to simulate 1) the movement of water (hydrodynamic or fluid transport), 2) the transport of particulate matter (sediment transport), and 3) the chemical transfers and kinetics, including water column–sediment exchanges. However, depending on the model being used and the experience of the analyst, optional stand-alone hydrodynamic, sediment transport, and chemical equilibrium models (upper portion of Figure 2-1), may also be used to evaluate the fate and transport model inputs. The output of the chemical fate and transport model is a characterization of the exposure concentrations, over time and space, of the constituent of interest. These concentrations are used in conjunction with one or more biological models (bottom of Figure 2-1) to evaluate bioaccumulation and toxicity of the chemical of interest to target organisms. Jørgensen (1990) described the development of a general model framework for metals in aquatic ecosystems that is very similar to the types of models covered in this review. Five submodels were discussed, including a submodel that describes the distribution pattern of metals, as well as speciation, sediment flux, bioaccumulation, and effects submodels.

Figure 2-2 provides a slightly more detailed representation of the important fate and transport model components and how the model results are used to evaluate biological effects. The sequence of steps followed when conducting a fate and transport analysis typically proceeds from left to right on this diagram, regardless of the model or models employed. The fluid or hydrodynamic transport analysis, performed using the submodel shown on the left, is completed to characterize the fluid transport regime, or movement of water, in the system. This movement of water is important because it provides dilution and mixing of the chemical in the receiving water. It also defines the relevant timescale, via its relationship to residence time, over which reactions may occur. As illustrated to the right of the hydrodynamic submodel on Figure 2-2, the water movement also affects the sediment transport in the water body. It does this by inducing shear stresses that control the settling and resuspension (scour) of particulate material between the water column and the bed. Particulate transport is frequently an important fate-controlling process because of the tendency of many organic chemicals and metals to sorb to particles and then be transported along with the particles. Hence, proper evaluation of the movement of the particles, especially between the water column and bed sediment, is an essential aspect of the chemical fate and transport analysis.

The evaluation of chemical reactions and transfers is also critical to the evaluation of exposure levels in aquatic settings. The third section of Figure 2-2 provides a

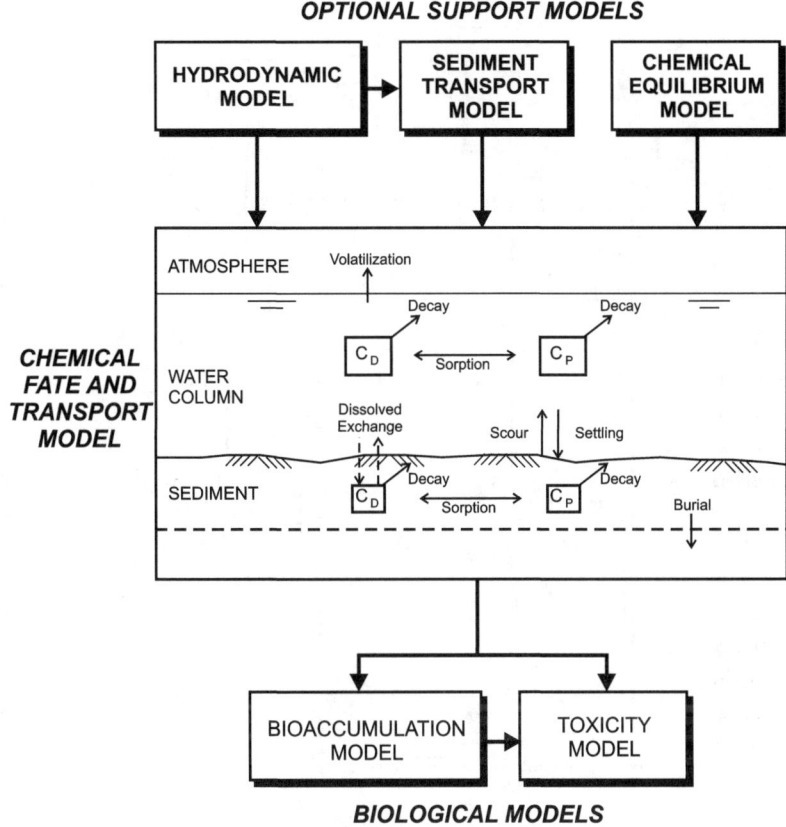

Figure 2-1 General fate and transport model framework

relatively simple representation of the complex set of reactions and transfer processes that may occur. Equilibrium partitioning of the chemical between the water (dissolved chemical) and the particles (sorbed or particulate chemical) is typically assumed to occur in both the water column and bed sediment in the simpler models, with the magnitude of the partition coefficient often evaluated in terms of the organic carbon content of the particles. Complexation of the chemical with dissolved organic matter (DOM), often expressed in terms of dissolved organic carbon (DOC), also is considered in a similar manner in some models. These partitioning reactions are important for a number of reasons. First, the sorbed chemical is transferred between the water column and sediment in association with the settling and resuspension of particulate matter, or in some models, via bulk exchange of water and sediment. Also, because the partition coefficient sets the fraction of total chemical that is in the dissolved form, and the diffusive flux between the water column and sediment pore water is proportional to the total dissolved (free + DOC-complexed) chemical concentration gradient between these

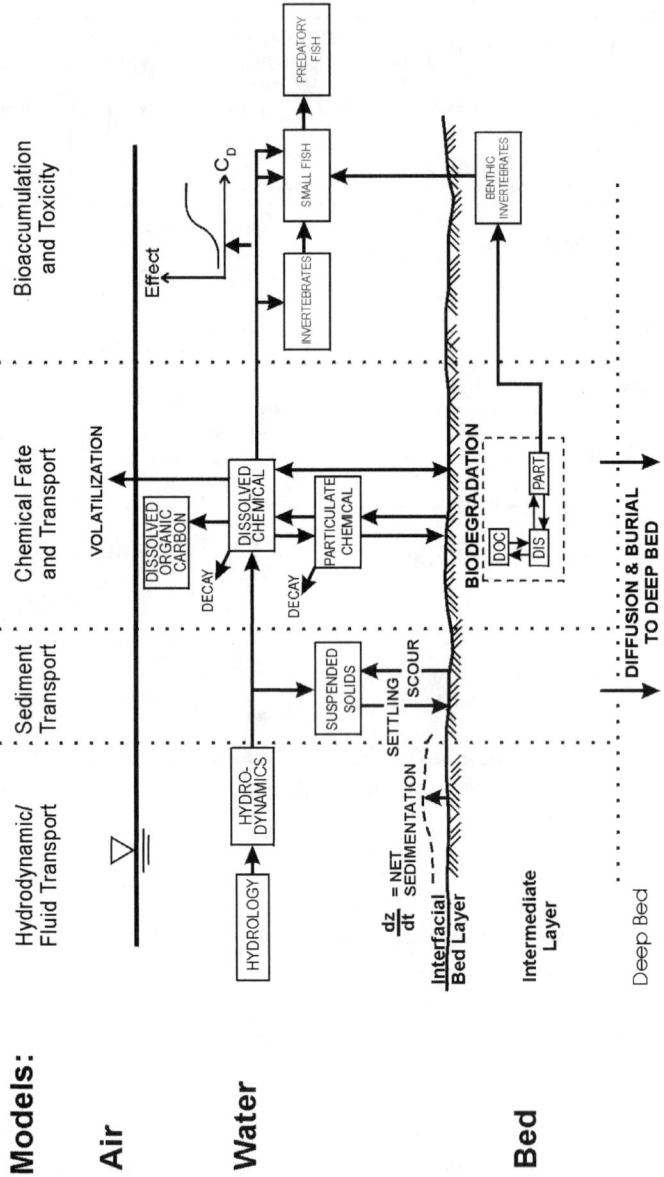

Figure 2-2 Example modeling framework

compartments, it affects this transfer process as well. Finally, partitioning is important because it often is assumed that the chemical sorbed to particles or complexed to DOC is not bioavailable (Landrum et al. 1985; Black and McCarthy 1988; USEPA 1998a). When this is the case, this fraction would not contribute directly to accumulation or toxicity. Alternatively, should the dietary route of exposure be deemed to be important, then sorption to particles that are subsequently ingested could become an important route of exposure.

The chemical reactions and transfers shown on Figure 2-2 are representative of the principal mechanisms included in most fate and transport models developed for organic chemicals. Lumped first-order decay processes, although important for many organic compounds, are generally of limited importance for metals. In some relatively simple models, or in screening-level analyses, first-order decay is sometimes used to represent net removal of sorbed metal from the water column by settling of particulate or sorbed metal. While this approach may be applied to metals, it typically provides an overly simplistic representation of the reactions and transfers of metals that take place in aquatic systems.

Relatively complicated organic and inorganic speciation and complexation reactions and competitive sorption of charged metal species to particles occur in the case of most metals, but these types of processes have only been included in fate and transport models in a few instances (Chapman 1982; Felmy, Brown et al. 1984a; EPRI 1996). Metal speciation computations are also needed in the context of the Biotic Ligand Model (BLM), a model being developed to evaluate metal bioavailability and toxicity (Di Toro et al. 1997, 2001; Paquin et al. 1998, 1999; Santore et al. 1998, 2001; Meyer et al. 1999; USEPA 1999a). Given the importance of these advanced computations in the assessment of metal fate and bioavailability, and in the evaluation of a predicted environmental concentration (PEC), relatively advanced chemical equilibrium models will be considered in detail as part of this review (Chapter 4).

It should be recognized that modeling evaluations of the fate and transport of metals such as mercury and selenium become increasingly complicated because these metals are subject to methylation and demethylation reactions and because some forms have a relatively high vapor pressure, making volatilization an important consideration (Masscheleyn and Patrick 1993; Zillioux et al. 1993). More generally, fate and transport models developed for specific metals that exist as organometallic forms and/or undergo changes in redox state (e.g., arsenic, chromium, mercury, and selenium) are not readily applicable to metals such as aluminum, cadmium, copper, nickel, lead, silver, and zinc. As a result, relatively specialized metal-specific fate and transport models developed for these types of metals (e.g., Mercury Cycling Model [MCM]; Hudson et al. 1994) are not considered in detail in this review. It is noted, however, that some of the more generalized models reviewed herein, such as Water Quality Analysis Simulation Program, Version 5 (WASP5) or Exposure Analysis Modeling System (EXAMS), although not designed for this purpose, can be configured to handle forward and back transformation reactions (methylation and demethylation or oxidation and reduction).

In addition to being used in conjunction with the predicted no-effect concentration (PNEC), the PECs that result from the fate and transport analysis also may be used in the context of a biological model framework, shown on the right side of Figure 2-2. The exposure concentrations in both the water column and sediment are used to evaluate both bioaccumulation and toxicity. Bioaccumulation is evaluated in the context of a food-chain model, of which several will be considered herein. Predictions of levels of metals that bioaccumulate in the tissues of aquatic life are needed to evaluate the dietary exposure of higher trophic levels, including both wildlife and humans. Further, as the understanding of how metals exert toxicity continues to be advanced, it is envisioned that it will be possible to relate chronic effects to critical body residue (CBR) levels or to organ-specific body burdens. Ultimately, it also may be possible to use tissue accumulation levels to assess the potential for effects on organism and population growth.

With regard to assessing the potential for toxic effects, the approach is to make a direct comparison of the PEC to the PNEC. The PNEC is typically set equal to the applicable acute or chronic water-quality criterion (WQC), assuming that the requisite WQC have been developed. Recent advances for metals have led to an improved understanding of how to evaluate metal bioavailability and of the mechanisms by which the acute toxicity of metals is exerted. As a result, the BLM framework is being developed and tested (USEPA 1999a). Initial results, primarily for copper and silver, but believed to be indicative of the potential utility of the BLM for other metals as well, indicate that this approach will provide an improved method for predicting the acute toxicity of metals over a range of water quality characteristics. The applicability of this approach to the evaluation of chronic toxicity is being evaluated at this time. The current status of the BLM will be discussed as part of the review of bioaccumulation and toxicity models presented in Chapter 5.

Another significant area where further model development is required involves metals in sediments. For example, models that simulate simultaneously extracted metal (SEM) and acid-volatile sulfide (AVS), factors shown to be closely related to metal toxicity in sediments (Di Toro et al. 1992; Ankley et al. 1993, 1994; Ankley, Di Toro et al. 1996), have received little attention to date. This type of model is introduced in Chapter 5 and discussed further in Chapter 6. Related processes, such as the oxidation of reduced metal sulfides, also have an important bearing on the form and bioavailability of metals, particularly in aerobic bottom sediments, and will control the fluxes of metals between the water column and sediment as well. Models of these processes are currently under development in conjunction with ongoing research and development efforts (Di Toro, Mahony, Gonzalez 1996; Di Toro, Mahony, Hansen, Berry 1996) but are only beginning to be incorporated in fate and transport models.

Scope of Literature Review

Although, in principle, relatively sophisticated stand-alone models (i.e., hydrodynamic, sediment transport, and chemical equilibrium models) could be incorporated in a unified fate and transport model framework, this is typically not the case. As a general rule, the stand-alone models that might be used to generate inputs for fate and transport models tend to incorporate a greater level of detail and sophistication than the currently available unified fate and transport models themselves. It will be seen that an important way that fate and transport models differ is the degree to which they have integrated some of the more refined features included in the stand-alone models. For this review, attention will be focused on the unified fate and transport models rather than the stand-alone hydrodynamic and sediment transport models that may be used to evaluate model inputs. However, for metals, the chemistry computations are of critical importance. Hence, particular attention will be given to this aspect of the fate and transport models. The review will also focus on the biological models and methods used to predict bioaccumulation and toxicity, since this capability is frequently required in the completion of an aquatic risk assessment.

A detailed description of example model applications, including model calibration, verification, and projection results, is not included herein. However, the interested reader is encouraged to refer to the numerous examples of model applications that are cited throughout the text because these references will frequently present such results. Care should be taken in considering such comparisons of model and data for purposes of evaluating the applicability of a particular model, however, because the ability of a model to reproduce a given data set is dependent on a number of factors other than the validity of the model framework itself. These other factors include but are not limited to the quality of the data, the consistency of field survey conditions with model assumptions (e.g., steady-state flows and loads), and the level of experience of the analyst. In view of these complicating factors, a determination of the adequacy of the fit of the model results to the data, within the context of regulatory and risk assessment goals, is a major challenge of any modeling effort.

Examples of field test results of some models will be discussed, as appropriate, throughout this review. In this regard, Dzombak and Ali (1993) provide an excellent review of hydrochemical modeling of metal fate and transport in freshwater environments. Tischler (1998) reviews the use of fate and transport models in waste load allocations for metals and discusses several case histories. However, actual testing and benchmarking of the models reviewed herein was beyond the scope of this literature review.

CHAPTER 3

Review of Aquatic Fate and Transport Models

The review of fate and transport models was directed at models that are applicable to metals and metal compounds. However, with only a few exceptions, most of the more widely used fate and transport models were originally developed for application to neutral organic compounds. As a result, they do not typically incorporate the sophisticated and useful inorganic chemistry algorithms that are available in stand-alone chemical equilibrium models. Even so, these fate and transport models have historically been applied to the analysis of metals, and though they do not provide as detailed a description of the important metal-related fate processes as would be desired, they still serve as useful frameworks for analysis purposes. As such, the review of fate and transport models will include those models developed for either organic chemicals or metals.

This chapter begins with a description of the search strategy used in the literature review ("Literature Search Strategy," p 14). A discussion of each of the models selected for inclusion in the review follows ("Review of Models," p 14). To highlight how the models have gradually increased in sophistication over time, the initial discussion follows the common lines of development along which some of the models have evolved. Probabilistic considerations also are discussed briefly. The key features of these fate and transport models are then described and compared in detail, and the advantages of alternative approaches are more critically discussed. Comparisons are made of the types of water bodies and fluid transport regimes to which the models are applicable, and of the time scale, space scale, and dimensionality of each model. The various approaches used to represent particulate transport and the ways in which chemical partitioning, reactions, and transfers are formulated also are discussed, with the advantages and disadvantages of alternative model formulations noted. Availability of the models and of technical support (based on information available at the time of this review) and the level of experience required to use these models are also indicated. This chapter concludes with a discussion of additional fate and transport model reviews and technical guidance documents that are available as supplemental references ("Summary of Model Reviews and Guidance Documents," p 51).

Literature Search Strategy

The literature search was conducted as follows. First, Current Contents on Diskette (CCOD), a subscription database service, and DIALOG, an extensive online database, were searched for citations and/or abstracts related to topics of interest (e.g., using combinations of key words such as "fate and transport," "metal," "model," etc.). Titles and abstracts were downloaded and reviewed to identify relevant articles. The references that appeared to be relevant were then ordered for use in a more detailed review. Due to the limited 3-month time frame for completion of the initial literature review, selections were also downloaded from Internet web sites, photocopied from readily accessible public and university libraries, or "rush" ordered through library services. HydroQual's in-house library and other technical files were also reviewed for relevant papers, reports, and models and their related documentation. These sources included references from the non-peer reviewed "gray literature."

References were next examined to identify models that could be used in exposure assessments for organic chemicals, metals, and metallic compounds in surface waters and sediments. All papers, reports, user guides, and other model-related documents that were retrieved were critically evaluated with respect to the applicability of the individual models to fate and transport analyses in general, and to analyses for metals in particular. A similar approach was followed for the literature searches conducted for both the chemical equilibrium models (Chapter 4) and the bioaccumulation and toxicity models (Chapter 5).

It is possible that some of the models that are described herein are no longer generally available, or that the original developers no longer provide technical support for a particular model. However, the cited documents are still of value for individuals who are currently developing models or who wish to learn more about alternative modeling approaches. The appendix to this book provides a partial list of sources of some of the models that are reviewed in this document.

Review of Models

This section provides a relatively detailed discussion of the models that have been reviewed. It begins with a description and overview of the various models. This is followed by a detailed comparison of model features.

Overview and description of models

The literature review led to the identification of numerous fate and transport models, most having been developed over the preceding 20 years. Although they vary markedly in degree of sophistication, the models that were reviewed generally include the basic components discussed previously, that is, fluid transport, sediment transport, and chemical reaction and transfer algorithms. With only limited exceptions, the models reviewed do not incorporate integrated bioaccumulation or toxicity subroutines. Rather, the predicted exposure levels are either entered into a separate bioaccumulation model, or they are compared to effect levels or water-quality criteria (WQC) for purposes of evaluating risk to the biota.

The fate and transport models included in this review are listed in Table 3-1. The table includes the acronym for each model, the full model name, and the citation where additional information about the model is available. Many of these models were originally developed during the late 1970s or early 1980s. They either continued to be refined following their initial release or publication, or they led to the development of new models that were essentially variations of the predecessor models. As a result, although 23 models are included in Table 3-1, there are evolutionary lines of development that result in several distinct groups of models that are interrelated, to a greater or lesser degree. Recognition of this common lineage is made in the discussion that follows. This will make it easier to understand the differences that exist among related models within a given group and will reduce the complexity of the overall review.

The models selected for inclusion herein were chosen because they
- have relatively widespread acceptance and frequently have been used by the scientific and regulatory community,
- incorporate features that make them well suited for specific applications, or
- incorporate features that are of particular interest with regard to modeling metal fate and transport.

The models represent a wide range of levels of complexity, with the simpler models best suited for use in screening-level analyses. These models are also best suited for use by inexperienced modelers. It should be emphasized that it was not possible for this review to be all inclusive, and it should not be considered as such.

The remainder of this section presents a general description of the important features of each of the models in Table 3-1. Whenever possible, the discussion will trace the common lines of development of related models or refer to a previously discussed predecessor model. This approach will highlight the changes that were made as the models evolved over time. The first 8 models in Table 3-1, analytical solution models, will be reviewed first. Next, 4 steady-state numerical solution models will be described and then 11 time-variable numerical solution models. The next section, "Comparison of fate and transport model features" (p 35), will also compare these same models, but will be organized to highlight, across all 23 models, differences in key model features.

Analytical solution models

The first 8 models of Table 3-1 are classified as analytical solution models. That is, the governing differential equation for conservation of mass is solved analytically to yield a continuous solution of concentration over time or in space. These

Table 3-1 List of fate and transport models

Acronym	Model name	Reference
Analytical solution models		
WQAM	Water Quality Assessment Methodology	Mills et al. 1982a, 1982b, 1985
RIVRISK	River Risk	Grieb 1995; EPRI 1996
SLSA	Simplified Lake and Stream Analysis	Di Toro, O'Connor et al. 1981; HydroQual 1982a
GMIII	General Motors III	HydroQual 1982b; Ogden 1984
MICHRIV	Michigan River	Delos et al. 1984; Mills et al. 1985
DJOC	Donald J. O'Connor (1988), a triad of papers	O'Connor 1988a, 1988b, 1988c
QWASI	Quantitative Water Air Sediment Interaction	Mackay et al. 1983; Mackay 1991
USES	Uniform System for Evaluation of Substances (replaced by EUSES; EC 1996b)	RIVM et al. 1994, as described in Johnson and Luttik 1995
Steady-state numerical solution models		
CTAP	Chemical Transport and Analysis Program	HydroQual 1981, 1982a
PAWTOXIC	Pawtuxent Toxics	Wright 1987
SMPTOX3	Simplified Method Program-Variable-Complexity Stream Toxics	LTI 1992
	Version 2	Dilks et al. 1994, 1995
MEXAMS	Metals Exposure Analysis Modeling System Includes: EXAMS MINTEQ	Felmy, Brown et al. 1984 Burns et al. 1982 Felmy, Girvin, Jenne et al. 1984
Time-variable numerical solution models		
EXAMSII	Exposure Analysis Modeling System-II	Burns and Cline 1985 (Also, Burns et al. 1982; Burns 1990, 1997)
RIVEQLII	River Quality II	Chapman 1982
WASTOX	Water Quality Analysis Simulation of Toxics	Connolly and Winfield 1984
RCATOX (AESOP)	Row Column AESOP for Toxics (AESOP: Advanced Ecological Systems Operating Program)	HydroQual 2003
WASP5	Water Quality Analysis Simulation Program, Version 5	Ambrose et al. 1993
	Distributed with DYNHYD5, Dynamic Hydrodynamics 5	Ambrose et al. 1993
DELFT3D	Delft 3D Model	Delft Hydraulics 1998
HSPF	Hydrologic Simulation Program-FORTRAN	Donigian et al. 1984; Bicknell et al. 1993
CHNTRN	Channel Transport Model	Yeh 1981, 1982, as described in USEPA 1987
FETRA	Sediment/Radionuclide Transport Model	Onishi and Thompson 1984, as described in USEPA 1987
SERATRA	Sediment Contaminant (i.e., radionuclide) Transport	Onishi and Wise 1982a, 1982b
RECOVERY	RECOVERY	Boyer et al. 1994; Ruiz et al. 2000

Note: Several additional models that had not been identified for purposes of the initial detailed model review, but have since been included via brief description in Chapter 3, are as follows: 2 analytical solution models, EUSES, SYVAC, TRANSPEC/BIOTRANSPEC, META4, San Francisco Bay model (Chen 1996), and MIKE21.

models tend to be relatively simple to apply and have low computational requirements. The important features of the 8 analytical solution models are compared in Table 3-2. The model names are indicated at the top of each column heading.

The Water Quality Assessment Methodology (WQAM) (Mills et al. 1982a, 1982b, 1985) is listed here as the first analytical solution model. It is actually a collection of screening-level analyses that can be applied in water quality assessments for toxic and conventional pollutants. It is included here because it provides an appropriate place to begin a water quality analysis, especially one performed by a relatively inexperienced modeler. Both of these documents include collections of formulas, tables, and graphs for use in preliminary assessments and computations that can be performed with a handheld scientific calculator. Although not specifically directed at evaluations for metals (the more recent 1985 publication does devote a section to metals), these guidance documents describe procedures for making relatively simple model-oriented calculations that are useful in a preliminary assessment of metal fate and transport. Lakes, rivers, and streams; impoundments; and estuarine settings are discussed in both documents. The 1985 publication also includes procedures for analyzing groundwater settings.

The second analytical solution model, River Risk (RIVRISK), is significantly different from the other analytical solution models that are reviewed. RIVRISK is a Windows-based program that calculates dissolved chemical concentrations using a 2-dimensional (2-D) steady-state analytical solution of the advection-dispersion equation. This vertically integrated model, applicable to rivers and streams, was developed for the Electric Power Research Institute (EPRI 1996) for use by member companies. The model uses the predicted exposure levels to assess carcinogenic and noncarcinogenic risks to human health. Although formulated relatively simply, using bioconcentration factors (as distinct from bioaccumulation factors), RIVRISK is the only fate and transport model in this review that also includes an integrated bioaccumulation submodel. (Chapter 5 includes a review of several state-of-the-art bioaccumulation models that are more suitable for use in an advanced analysis of bioaccumulation.)

RIVRISK considers the impacts of multiple point source loads by superposition of the solutions for individual sources. The model includes net settling of suspended solids from the water column to the bed, but does not account for resuspension from the bed (Grieb 1995). It also computes the diffusive flux of dissolved chemical between the water column and sediment pore water. Chemical speciation information is included via an online database (EPRI 1996). The database contains speciation results for a number of metals based on water quality characteristics in 19 rivers from across the U.S. From the database, the user selects the river having characteristics that most closely resemble the receiving water of interest. The speciation information that was previously saved in the database is then used to calculate the important dissolved species and to estimate the solubility of the metal under consideration. This approach is comparable to use of a separate chemical

Table 3-2 Summary of features of analytical solution models

Characteristics and processes	WQAM	RIVRISK	SLSA	GMIII	MICHRIV	DJOC	QWASI	USES[a]
Water body								
River or stream	X	X	X		X	X	X	
Lake or reservoir	X			X			X	X
Estuary	X							
Fluid transport								
User specified or defined	X	X	X	X	X	X	X	X
Hydrodynamic submodel								
Time dependence								
Steady state	X	X	X	X	X	X	X	X
Dynamic				X			X	X
Dimensionality								
1D	X	X	X	CM[b]	X	X	CM	CM	CM
2D (2H or 2V limitation)		2H[c]						
3D								
Bed layers	1	1	1	1	1	1	1	1
Particulate transport								
Net settling (S) or resuspension (R)								
Settling	X	X	X	X	X	X	X	
Resuspension	X		X	X	X	X	X	
Bulk exchange								
Bioturbation								
Scour or burial	X		X	X	X	X	X	
Expanding bed or storage								
Particle types	1	1	1	1	1	1	1	1
Chemical reactions and transfer								
Equilibrium (E) or nonequilibrium (N) sorption to particles	E	E	E	E	E	E	E	E
DOC complexation								
Water-bed diffusive exchange	X	X	X	X	X	X	X	X
Inorganic speciation or complexation		X						
Precipitation or dissolution		X						
SEM/AVS								
Degradation or transformation process								
First-order decay	X	X	X	X	X	X	X	X
Photolysis	X	X						
Hydrolysis	X	X						
Oxidation	X	X						
Volatilization	X	X				X		X
Daughter product								
Availability								
Public domain	X		X		X	X	X	X
Proprietary		X		X				
Technical support								
USEPA	X				X			
Non USEPA		X	X	X		X	?[d]	?

[a] EUSES current version (EC 1996), features may differ; [b] CM = completely mixing volume; [c] 2H = 2-D horizontal plane; [d] ? = unknown.

equilibrium model to predict speciation reactions, with the advantage that the information is internally specified. The disadvantage is that a relatively limited number of water quality data sets are available to characterize the water body being analyzed.

RIVRISK has been applied by a number of electric power utilities. Example applications include an evaluation of a dioxin-containing discharge into a small river, thermal impacts on a cold-water fishery, and impacts to a river resulting from underground seepage of an iron-cyanide complex (EPRI 1996). The proprietary software is owned by EPRI, and at the time of this review it was available only to member companies. Technical support is available to registered users upon referral by EPRI.

The next 4 analytical solution models are all variations of the Simplified Lake and Stream Analysis (SLSA) Program. SLSA was originally developed for the Chemical Manufacturers Association (CMA) (Di Toro, O'Connor et al. 1981; HydroQual 1982a; Di Toro 1987). It is applicable to steady-state conditions in a 1-dimensional (1-D) stream having constant geometry and in a completely mixed lake. The model includes an option to compute the water column response to depuration of a contaminated sediment, once the load has been discontinued. SLSA may also be applied in time-variable mode to a lake. It includes a single particle-size class, particle settling and resuspension, diffusive transport of dissolved chemical between the water column and an interactive bed layer, and first-order decay. Although applicable to metals, simple, instantaneous, linear equilibrium partitioning between the dissolved and particulate fractions is assumed. Suspended solids are also constant within the river reach. The SLSA equations have been applied successfully in time-variable analyses of polychlorinated biphenyls (PCBs) in the Great Lakes (Thomann and Di Toro 1983) and to DDE (dichlorodiphenyldichloroethylene) and lindane in a quarry (Di Toro and Paquin 1984). SLSA was also applied in a side-by-side test with the Exposure Analysis Modeling System (EXAMS) in the analysis of endothall, an aquatic herbicide, in an experimental lake (Reinert and Rodgers 1986). The assessment of the authors that neither model was successfully validated in this latter test was attributed to uncertainty in fate processes such as the biotransformation rate of endothall. An advantage of SLSA is that the input requirements are relatively simple, making it easy to apply in screening-level analyses. However, the geometric configurations and loading scenarios that can be simulated are relatively simple as well, and thus, it may not be readily applicable to a moderately complex setting.

General Motors III (GMIII), an adaptation of SLSA, was developed to overcome SLSA's inability to represent variable channel geometry (HydroQual 1982b). It is a 1-D model that is applicable to streams, but has the capability of representing multiple constant geometry river reaches, with each having a unique set of cross-section, flow, and particulate transport parameters. GMIII has been applied to the analysis of suspended solids, nickel, and zinc in the Clinton River (HydroQual 1982b) and to zinc in the Saginaw River (Ogden 1984; HydroQual 1989).

The Michigan River (MICHRIV) model is a steady-state, 1-D, analytical solution model that is applicable to streams and rivers (Delos et al. 1984). It is also derived from SLSA and has essentially the same requirements and capabilities as GMIII. In particular, it can be configured for multiple constant geometry reaches. It includes a lumped first-order decay term. The model predicts a spatially varying water column suspended solids concentration (one particle class) and total and dissolved chemical concentrations in the water column and sediment. Mills et al. (1985) provide a description of this model and also present a simplified version of the analytical solution that can be applied if the bed sediment chemical concentration is specified by the analyst. This assumption is intended for use in analyzing situations where net scour of the bed occurs over a relatively short period of interest and where the water column chemical concentration increases in the downstream direction. Of the 3 preceding SLSA-based models, SLSA is available through CMA, and MICHRIV is available through the U.S. Environmental Protection Agency (USEPA). It is not known at this time if GMIII is currently available for use.

The next analytical solution model in Table 3-2, DJOC[1], has not actually been programmed as a computer model. Rather, the name refers to a series of 3 papers that present analytical solutions for chemical fate and transport in lakes and reservoirs and in a 1-D river (O'Connor 1988a, 1988b, 1988c). The analytical solutions incorporate refinements to SLSA, such as allowance for a variable suspended solids concentration in a constant geometry reach. Although the solutions are not available in the form of a computer program, this series of papers is included in this review because they provide an excellent description of the equations that define the steady-state distribution of solids and sorptive chemicals in freshwater systems. The essence of the discussion on model development that is presented in these papers applies to many of the models that are reviewed here. The first paper, in particular, includes an excellent description of the development and formulation of the governing differential equations in a typical fate and transport model. The interested reader desiring a more formal introduction to the underlying principles and governing equations of fate and transport models than is presented herein is referred to this series of papers.

The next analytical solution model to be reviewed is the Quantitative Water Air Sediment Interaction (QWASI) model that is applicable to lakes and rivers (Mackay, Paterson, Joy et al. 1983; Mackay, Joy, Paterson 1983). This model was designed to simulate the exchange of chemicals with a measurable vapor pressure between the atmosphere, water column, and benthic sediment based on the thermodynamic principle of fugacity (Mackay and Paterson 1981, 1982). The

---

[1] The acronym DJOC, based on the initials of the model developer, Donald J. O'Connor, is adopted for use in this review.

QWASI model was subsequently adapted to consider metals through a simple numerical manipulation of fugacity (Mackay and Diamond 1989; Diamond et al. 1990; Mackay 1991). The adapted model uses "equivalent equilibrium aqueous phase" concentrations, or "aquivalence," as an alternative equilibrium criterion for chemicals with or without a measurable vapor pressure. Losses from the model are burial, volatilization to an infinite air compartment, and transformation in air, water, and sediment in the case of degradable chemicals.

Recognizing that the aquivalence-based QWASI model assumes simple equilibrium partitioning of a single species, Diamond et al. (1992) further modified the model to consider multiple interconverting species as occurs with metals. The model requires specification of the fractions of each chemical species in air, water, and sediment and the partition coefficient or total chemical in water and sediment. The mass balance equations are solved numerically for total metal in each compartment, from which concentrations and transport and interconversion rates are calculated for each species. The model includes air as a compartment with infinite volume in order to accommodate air–water exchange of volatile species such as elemental mercury and some organometallics (e.g., monomethyl mercury).

Although not available to be considered in detail at the time of this review, the multispecies QWASI model has most recently been adapted to consider speciation and complexation of metals. This has been achieved by its loose coupling to MINEQL+ (Bhavsar et al. submitted). The TRANSport and SPECiation model, TRANSPEC, considers metal in dissolved, colloidal (sorbed to dissolved organic carbon [DOC]), and particulate phases in water and a 2-layer sediment system. Air is also considered in the case of mercury. The model is capable of estimating species fractions and the partition coefficient in the water column. Limitations in the speciation model in terms of metal partitioning to solid phases now preclude its use to estimate the metal partition coefficient in the sediments. While the speciation model assumes equilibrium, the fate and transport component can be run in steady- or unsteady-state modes. The model is run by first computing metal speciation for the range of scenarios specified according to ambient chemistry data. The transport module then runs by accessing species fractions and water column partition coefficients. The model has been evaluated by application to a lake with elevated concentrations of zinc in the sediments and the further complication of seasonal redox variations in the sediments causing variations in porewater concentration in the diffusive flux between sediment and water. A further adaptation of the model, BIOTRANSPEC, has been applied to consider mercury dynamics (Gandhi et al. in prep). This version of the model considers the speciation of dissolved mercury species, redox-driven dynamics of elemental mercury, and kinetically dependent methylation–demethylation reactions (the latter relies on site-specific rate constants). The resultant species fractions serve as input values to the transport module that also includes methylmercury accumulation by fish.

The last of the 8 analytical solution models to be reviewed is the Uniform System for the Evaluation of Substances (USES) (RIVM et al. 1994, as described by Johnson and Luttik 1995). As applied in this risk assessment for antifoulants (i.e., irgarol, diuron, tributyltin, and copper), where a middle size yacht basin is modeled, USES consists of a series of relatively simple calculations for a completely mixed water body. For this particular model application, the leaching rate of the antifoulant from boat hulls, the number of boats in the basin, the boat hull surface area, and other relevant information needed to evaluate the loading rate to the basin also are specified. The basin volume and hydraulic residence time are evaluated. The total dissolved concentration in the water column is then estimated from the loading and residence time information, assuming simple equilibrium partitioning. (For degradable compounds, a first-order removal rate coefficient may also be assigned.) The authors also present a simplified partitioning-based approach for predicting the concentration of an organic substance in the sediment. Given the complexities associated with modeling metals in sediments, the authors recommend that the risk characterization for sediments not be performed without having measured sediment data. The utility of the acid-volatile sulfide (AVS) and simultaneously extracted metal (SEM) approach for evaluating sediment toxicity due to metals, as described in Chapter 5, is emphasized. A method for evaluating a first-order "disappearance rate" of dissolved metal from the water column via diffusion to a sulfide-rich sediment layer is also presented.

Although the approach followed in this application of USES was quite simple, the credibility of this screening-level analysis was enhanced by the analysts' comparison of model results to monitoring data for 3 sites in Sweden and 2 sites in the Netherlands. Even with this relatively simple model, reasonably good agreement between predicted water column concentrations and measured data was achieved. It was not clear from the description of USES presented in this risk assessment whether this model is suitable for application to other types of settings.

The USES model has been superceded by a more complex version, the European Union System for the Evaluation of Substances (EUSES) software (EC 1996b). This updated European version of USES is a relatively complex model (Berding et al. 1999) that is based on the calculations and mathematical models presented in the European Commission's technical guidance document (TGD) (EC 1996a). As a late addition to this review, EUSES is not compared in detail in the accompanying tables.

The review of analytical solution models concludes with a discussion of the surfacewater submodel of the Systems Variability Analysis Code (SYVAC), a biosphere model that incorporates soil, atmosphere, and food-chain submodels as well. Although not included in the detailed comparison tables, this recent addition to this review is the surfacewater model that has been used to assess Canada's nuclear fuel waste disposal concept (Bird et al. 1993). It is a simple, time-dependent, mass balance, screening-level model of a lake that calculates radioactive and stable isotope concentrations in a completely mixed water column compartment. Losses include flushing from the lake, volatilization, radioactive decay (forming stable elements), and first-order transfer from the water column to the sediment, the second model compartment. Monte Carlo simulation techniques are used to predict output concentrations, with many of the model inputs assigned as lognormally distributed random variables. The resulting exposure concentrations are used in the other submodels to predict the dose to exposed biota and humans. Model validation tests indicate that the lake model predictions are consistent, within reasonable limits of uncertainty, with empirical data for a number of Canadian lakes.

Steady-state numerical solution models

The rest of the models to be reviewed in Table 3-1 are numerical solution models. Of these, the first 4 are steady-state models and the final 11 are time-variable models. An important advantage of numerical solution models is that they can be used to represent more complex geometric configurations and transport regimes than analytical solution models. They are thus more readily applicable to multidimensional settings.

The important features of the 4 steady-state numerical solution models are compared in Table 3-3. Again, the model names are indicated at the top of each column heading. The first model is the Chemical Transport and Analysis Program (CTAP) (HydroQual 1981, 1982a). CTAP is simply the steady-state numerical solution version of SLSA. As such, it incorporates most of the same features as SLSA. Additionally, it provides much greater flexibility in representing complex system geometries, being suitable for use in 1-D, 2-D, or 3-D analyses. CTAP is also capable of simulating 5 classes of particles, a feature that is useful in representing particles types having different sizes or organic content. CTAP has been applied in a number of settings, including the development of total maximum daily loads (TMDLs) for 8 metals (arsenic, copper, cadmium, lead, mercury, nickel, zinc, and silver) in New York Harbor, USA (HydroQual 1995) and in the analysis of tributyltin oxide (TBTO) levels resulting from leaching from ship hulls in New York Harbor and Chesapeake Bay, USA (St. John et al. 1985).

Pawtuxent Toxics (PAWTOXIC) (Wright 1987), an adaptation of the 1-D steady-state stream water quality model QUAL-II (Brown and Barnwell 1985), is applicable to rivers and streams. It is a relatively simple fate and transport modeling framework that allows the user to specify a condition of either net settling or net resuspension of particles. Empirical relationships are used to evaluate the settling and resuspension rates. Equilibrium partitioning is assumed. The model does not include either bulk exchange of sediment (water + particles) or diffusive flux of dissolved chemical between the water column and bed. Thus, it provides one of the simpler representations of water column–bed interactions of the models reviewed. PAWTOXIC has been applied to the analysis of metals in the Blackstone River

Table 3-3 Summary of features of steady-state numerical solution models

Characteristics and processes	CTAP	PAWTOXIC	SMPTOX3	MEXAMS
Water body				
River or stream	X	X	X	X
Lake or reservoir	X			X
Estuary	X			
Fluid transport				
User specified or defined	X	X	X	X
Hydrodynamic submodel				
Time dependence				
Steady-state	X	X	X	X
Dynamic				
Dimensionality				
1-D	X	X	X	X
2-D (2H or 2V limitation)	X			X
3-D	X			X
Bed layers	M[a]		1	M
Particulate transport				
Net settling (S) or resuspension (R)		S or R		S
Settling	X		X	X
Resuspension	X		X	X
Bulk exchange				X
Bioturbation	X			X
Scour or burial	X		X	
Expanding bed or storage				
Particle types	5	1	1	1
Chemical reactions and transfer				
Equilibrium (E) or nonequilibrium (N) sorption to particles	E	E	E	E
DOC complexation				
Water-bed diffusive exchange	X		X	B[b]
Inorganic speciation or complexation				X
Precipitation or dissolution				X
SEM–AVS			V4[c]	
Degradation or transformation processes				
First-order decay	X		X	
Photolysis			X	
Hydrolysis				
Oxidation				
Volatilization			X	X
Daughter product				
Availability				
Public domain	X	X	X	X
Proprietary				
Technical support				
USEPA			X	X
Non USEPA	X	X		

[a] M = Multiple layers
[b] B = bulk exchange (particulate and dissolved chemical)
[c] V4 = SMPTOX4 only

(Massachusetts and Rhode Island, USA) (Wright et al. 1998; USEPA 1998b). Although the model successfully reproduced the spatial trends of total and dissolved metals (cadmium, chromium, copper, lead, and nickel) in the Blackstone River, the relatively simple approach used for particulate transport in this model could be a potential limitation in some situations.

The Simplified Method Program-Variable-Complexity Stream Toxics (SMPTOX) Model, SMPTOX3, is a Windows-based, 1-D, steady-state model that is applicable to rivers and streams (LTI 1992). This is another model that evolved from the earlier analytical solution models, along the SLSA-MICHRIV line of development. It is a numerical solution version that can be used to simulate multiple chemicals. It includes equilibrium partitioning and first-order decay (volatilization, hydrolysis, photolysis, and biodegradation). The theoretical basis of the model is described in Book 2, Chapter 3 of the USEPA guidance manual for waste load allocations (Delos et al. 1984). The model incorporates particle settling and resuspension and diffusive exchange of dissolved chemical between interactive water column–sediment layers. It predicts total suspended solids and total and dissolved chemical in the water column and bed. Although the model structure is similar in many respects to CTAP, a significant disadvantage is that it is only applicable to 1-D settings. An interesting feature of a developmental version of SMPTOX is that it incorporates AVS and SEM for the metals copper, cadmium, nickel, lead, and zinc (Dilks et al. 1994, 1995). As discussed in Chapter 5, AVS and SEM are important because they form the basis of the USEPA equilibrium partitioning sediment guidelines (ESGs) for metals. The seasonal variation in AVS, known to be a significant factor, cannot be predicted with this steady-state model. The model does not consider oxidation of metal sulfides either, a process that will affect the water column–sediment exchange of metal.

Metals Exposure Analysis Modeling Systems (MEXAMS) (Felmy, Brown et al. 1984) consists of 2 linked models, the Exposure Analysis Modeling System (EXAMS) (Burns et al. 1982), a fate and transport model, and MINTEQ (Felmy, Girvin, Jenne 1984), a chemical equilibrium model (discussed further in Chapter 4). As such, MEXAMS has the capability to simulate the fate and transport of metals using detailed chemical equilibrium calculations. Although the current version of EXAMS, EXAMSII, has time-variable capabilities and will be discussed later in this section, MEXAMS was developed using an earlier version that could only be used in steady-state mode. Thus, the earlier version of EXAMS is introduced here in conjunction with the discussion of MEXAMS.

EXAMS is a well-known and widely applied fate and transport model that was developed by the USEPA. It is capable of simulating 1-D, 2-D, and 3-D settings and includes an interactive bed layer. The stand-alone version of EXAMS assumes equilibrium partitioning. It also includes a relatively advanced set of subroutines to compute various degradation processes (hydrolysis, photolysis, volatilization, etc.) and to simulate the dissociation of ionizing substances. The MEXAMS incarnation of EXAMS transfers the total metal (arsenic, cadmium, copper, lead, nickel, silver, and zinc are included) concentrations computed by EXAMS to MINTEQ, where the distribution of metal species is computed. This information is then transferred back to EXAMS to continue with the fate and transport simulation. An iterative approach is applied because the speciation affects the fate and transport and hence the total metal concentration used by MINTEQ. The fate and transport module (EXAMS) uses bulk exchange between the water column and sediment rather than independent mechanisms of settling and resuspension of particles and sorbed chemical and diffusive exchange of dissolved chemical. This will be discussed in further detail later in this chapter ("Particulate transport mechanisms," p 42). Although MEXAMS has been refined to include net settling of particles to the bed, it does not maintain an accounting of this transfer with respect to sediment burial. Another limitation of MEXAMS is that, because it was developed to work with EXAMS rather than EXAMSII, its application is limited to steady-state analyses.

Medine and Bicknell (1987) applied MEXAMS to metals in the Naugatuck River in Connecticut. Dissolved organic carbon (DOC) was not considered. After calibration, the model yielded reasonably good agreement with total metal concentration data, but an assessment of its ability to predict metal speciation could not be made based on this analysis. MEXAMS was also applied to an analysis of Ten Mile River in Massachusetts and Rhode Island, USA (Medine and Bicknell 1987). The application was complicated by the fact that the underlying steady-state flow assumption in the early version of EXAMS was not applicable to conditions at the time of the survey. Total and dissolved copper were overpredicted in the water column and underpredicted in the sediment. This discrepancy may reflect the use of a bulk exchange coefficient to set the rate of water column–sediment exchange rates.

Dzombak and Ali (1993), in a review of metal fate and transport models, identified the assumption of steady-state hydraulic conditions and use of equilibrium conditions as major limitations in MEXAMS. They recommended that a kinetic formulation incorporating a pseudo-equilibrium approach, as applied in River Quality II (RIVEQLII) (Chapman 1982), be employed. The general absence of reactions to represent metal-DOC complexation and competitive sorption on solids was also cited as a significant limitation, not only in MEXAMS, but in all geochemical models available as of 1993.

Time-variable numerical solution models

The last 11 models listed in Table 3-1 are numerical solution models that are suitable for use in time-variable simulations of chemical fate and transport. It should be understood that, as a general rule, these time-variable models also may be used to simulate steady-state conditions. A steady-state analysis might be called for in a relatively simple waste load allocation situation where the regulatory agency is interested in evaluating water quality during critical low flow conditions. However, consideration should be given to the fact that the effort associated with setting up a time-variable model for a steady-state analysis is often significantly greater than the effort associated with use of a generally simpler steady-state model. Other considerations may control the decision about the type of model to use, however.

The important features of the 11 time-variable numerical solution models are compared in Table 3-4. A general description of each of these time-variable models follows.

As noted above, the original version of EXAMS, released in the early 1980s, was a steady-state model (Burns et al. 1982). Since then, EXAMS has continued to be refined, with USEPA providing ongoing technical support through the present time. An updated version, EXAMSII (Burns and Cline 1985), was released several years after the initial release of EXAMS. EXAMSII introduced several enhancements to the model, the most important being the ability to be run in both steady-state and time-variable modes. Version 2.97 is the most recent release of EXAMSII (Burns 1997). As will be discussed in the next section, a disadvantage of EXAMS and EXAMSII, and by inference of the metals version of this model MEXAMS, is that it represents water column–bed interactions by a bulk exchange of water and particles. This is judged to be an inherent limitation of EXAMS, one that could potentially limit its predictive capabilities, at least in some situations.

RIVEQLII (Chapman 1982) was developed for the purpose of simulating the transport of inorganic constituents such as metals. It incorporates relatively sophisticated chemical equilibrium calculations via a link with MINEQL (Westall et al. 1976), an advanced chemical equilibrium model (discussed in Chapter 4). RIVEQLII solves the 1-D advection–dispersion equation and is applicable to rivers and streams. It incorporates net, first-order settling of particles to the bed, sorption to reactive surfaces, and precipitation and dissolution of chemical to and from the bed. Equilibrium conditions generally are assumed at each time step for most reaction processes, with MINEQL providing the geochemistry results. An exception is that redissolution of precipitated metal is rate-limited by a time constant that allows only a fixed proportion of the precipitate that would dissolve under equilibrium conditions to undergo dissolution at each time step. Chapman (1982) applied RIVEQL to results of a field experiment where NaOH was added to Daylight Creek, a stream draining an abandoned mine located in New South Wales. The model achieved reasonable agreement with total and particulate sodium, zinc, aluminum, and copper over time and space.

The origins of the next 3 models to be discussed are rooted in the Water Quality Analysis Simulation Program (WASP) (Di Toro, Fitzpatrick, Thomann 1981) and in the chemical fate and transport framework incorporated in SLSA and CTAP, described previously. WASP was originally developed in the early 1970s as a proprietary water quality model. It was designed to be a flexible tool for simulating a variety of relatively complex time-variable water quality problems and, in particular, eutrophication. An updated public domain version was subsequently developed for USEPA (Di Toro, Fitzpatrick, Thomann 1981). The first model,

Table 3-4 Summary of features of time-variable numerical solution models

Characteristics and processes	EXAMSII	RIVEQLII	WASTOX	RCATOX	WASP5	DELFT3D	HSPF	CHNTRN	FETRA	SERATRA	RECOVERY
Water body											
River or stream	X	X	X	X	X	X	X	X	X	X	X
Lake or reservoir	X		X	X	X	X	X				X
Estuary			X	X	X	X		X	X		
Fluid transport											
User specified or defined	X	X	X	X	X	X	X	X	X	X	X
Hydrodynamic submodel			X	X	X	X					
Time dependence											
Steady-state	X	X	X	X	X	X	X	X	X	X	X
Dynamic	X	X	X	X	X	X	X	X	X	X	X
Dimensionality											
1-D	X	X	X	X	X	X	X	X	X	X	X
2-D (2H or 2V limitation)	X		X	X	X	X			2H[a]	2V[b]	
3-D	X		X	X	X	X					
Bed layers	M[c]	1	M	M	M	2	1	?[d]	?	M	M
Particulate transport											
Net settling (S) or resuspension (R)											
Settling		X	X	X	X	X	X	X	X	X	X
Resuspension			X	X	X	X	X	X	X	X	X
Bulk exchange	X										
Bioturbation	X		X	X	X	X					X
Scour or burial			C[e]	C	C	C	C	C	C	C	X
Expanding bed or storage			X							X	
Particle types	1	1	3	2	3	3	3	3	3	3	1
Chemical reactions and transfer											
Equilibrium (E) or nonequilibrium (N) sorption to particles	E		E	E	E	E	E,N	E	E	E,N	E
DOC complexation	X		X	X	X	X					
Water–bed diffusive exchange	B[f]		X	X	X	X	X	X	X	X	X
Inorganic speciation or complexation		X				X					
Precipitation or dissolution		X				X					
SEM–AVS											
Degradation or transformation processes											
First-order decay	X		X	X	X	X	X	X	X	X	X
Photolysis	X		X	X	X		X	X		X	
Hydrolysis	X		X	X	X		X	X		X	
Oxidation	X		X	X	X		X	X		X	
Volatilization	X		X	X	X	X	X	X		X	X
Daughter product	X			X	X		X		X		
Availability											
Public domain	X		X		X		X	X	X	X	X
Proprietary		?	X	X		X					
Technical support											
USEPA	X				X		X				
Non USEPA		?	X	X		X		?	?	?	X

[a]2H = 2-dimensional horizontal plane; [b]2V = 2-dimensional vertical plane; [c]M = multiple layers; [d]? = unknown; [e]C = calculated by program; [f]B = bulk exchange (dissolved and particulate chemical)

Water Quality Analysis Simulation of Toxics (WASTOX) (Connolly and Winfield 1984), is essentially a version of WASP with particulate and chemical transfers that are very similar to the transfers used in SLSA and CTAP. Since its initial development, WASTOX has undergone a number of refinements. A notable recent development has been conversion to a parallel processing version of WASTOX, the Row-Column AESOP for Toxics (RCATOX) model (HydroQual 2003). This model is applicable to multiple chemicals and simulation of reaction or daughter products.[2] Although currently in use, RCATOX is a developmental version that is still being tested and is not available for release at this time.

The third model, WASP5, is the current version of a flexible eutrophication and chemical fate and transport model supported by USEPA (Ambrose et al. 1993). Because of the variety of model names in circulation for what are essentially sequential releases of predecessor versions of WASP5, it is useful to briefly describe the nomenclature and historical development of the WASP family of models. The original version incorporating a chemical fate and transport subroutine was referred to as TOXIWASP (Ambrose et al. 1983). This was the name applied to WASP3 when it was initially linked with a chemical fate subroutine. This linkage and the features in the chemical fate model are described in the WASP3 User's Guide (Ambrose 1986). TOXIWASP incorporated the chemical reactions and transfers included in EXAMS into the generalized water quality model WASP, with some additional sediment transport capabilities. WASP3 was subsequently superseded by WASP4, which when linked with the chemical fate subroutine yields the chemical fate model TOXI4 (Ambrose 1988). It was the USEPA's intent that the WASP4 framework would supercede both WASTOX and TOXIWASP, by incorporating and expanding on aspects of each (Ambrose 1988). The transport structure and steady-state solution option of WASTOX was added, and a chemical fate package containing elements of WASTOX, TOXIWASP, and EXAMSII was also added. The WASP4 system also introduced an associated food chain model (Connolly and Thomann 1985). WASP5, the current version of this fate and transport model (Ambrose et al. 1993), is currently distributed with 2 different kinetic subroutines, EUTRO, for use in eutrophication problems, and TOXI5, for simulating chemical fate and transport. In summary, there are 2 models from this series of models that are still under development, WASP5 and WASTOX. These models, along with RCATOX, are reviewed herein.

An important feature that distinguishes WASTOX, RCATOX, and WASP5 from most of the other models in this review is that the current versions are designed to seamlessly interface with state-of-the-art hydrodynamic and/or sediment transport models. This is especially useful when modeling estuarine and coastal settings

---

[2] RCA, or Row-Column AESOP, is the parallel processing version of the Advanced Ecological Systems Operation Program. It was designed to take advantage of the parallel processing capabilities of modern-day supercomputers.

where complex hydrodynamic conditions often prevail. WASP5 is distributed by the USEPA with the hydrodynamic model DYNHYD5 (Ambrose et al. 1993), a link-node hydrodynamic model applicable to predominantly 1-D settings. It is PC-compatible and is distributed with pre- and post-processors for use with WASP5. WASTOX and RCATOX have the capability to readily interface with the state of the art Estuary, Coastal, Ocean Model (ECOM) family of hydrodynamic models (Blumberg and Mellor 1987), and with Estuary, Coastal, Ocean Model-Sediment (ECOMSED) (HydroQual 1998), a sediment transport model capable of simulating the movement of both cohesive and noncohesive sediments. The cohesive sediment transport model is based on the work of Lick and others (Ziegler and Lick 1986; Lick et al. 1987; Tsai and Lick 1987; Gailani et al. 1991).

Although not available at the time of this review, and hence not included in Table 3-4, the Environmental Fluid Dynamics Code (EFDC) is a relatively sophisticated hydrodynamic and sediment transport model that has recently been modified to simulate chemical fate as well. This model incorporates the basic theoretical framework of Blumberg and Mellor (1987), as implemented in the ECOM family of hydrodynamic models. A variety of cohesive and noncohesive sediment transport formulations (e.g., Lick et al. 1987; Garcia and Parker 1991) are available for modeling the suspended load. It also provides the capability to simulate bedload transport. EFDC also includes a sediment diagenesis model that was developed for Chesapeake Bay (Di Toro and Fitzpatrick 1993). The hydrodynamic and sediment transport components have been documented previously (Hamrick 1992, 1996), but the water quality documentation only recently has become available for review (TetraTech 2002a, 2002b). It is for this reason that EFDC is not included in Tables 3-1 or 3-4. When EFDC was applied to metals in the Blackstone River, simple linear partitioning was employed (Ji et al. 2002). Because it does not include an advanced chemistry subroutine, the model output must be used in conjunction with a separate chemical equilibrium model if a detailed evaluation of metal speciation is desired. Although envisioned for use in evaluation of discharge limits based on sediment metals, the version of this model currently under development does not explicitly consider AVS, SEM, or metal sulfide oxidation.

The Delft Hydraulics Hydraulic Lab, Delft, Netherlands, has developed DELFT3D, a flexible and integrated modeling environment that incorporates many of the fate and transport processes discussed herein (Delft Hydraulics 1998). DELFT3D is a sophisticated modeling package that incorporates hydrodynamic, water quality (including cohesive and noncohesive sediment transport), chemistry, and wave generation modules, among others. Although the chemistry module is described as "a general model," the input processor that is provided to model users is applicable only to certain "standard applications." DELFT3D is a proprietary software package with a commercially available version.

The next time-variable model listed in Table 3-4 is the Hydrologic Simulation Program-FORTRAN (HSPF) (Donigian et al. 1984; Bicknell et al. 1993). This model has a long history of development. Its origin is the Stanford Watershed Model (SWM), developed in the mid-1960s. It then became the Hydrocomp Simulation Program (HSP) in 1969 and the Pesticide Transport and Runoff (PTR) model in 1973. As one might infer from these titles, the model is unique among those reviewed herein in that it includes hydrology and nonpoint source terrestrial runoff modules that are perhaps more well known and widely applied than the fate and transport module that it also includes. This review will focus on the fate and transport component of HSPF.

HPSF is readily applicable to complex watersheds and was developed to provide continuous long-term and storm event simulations. It is used to model well-mixed reservoirs and rivers, including branched river systems. Water quality is simulated using lumped first-order decay, and first-order kinetic processes are used to model sorption to as many as 3 sediment types. The model has been used extensively throughout the U.S., frequently to model pesticides. It was also recently used to evaluate watershed contributions of flow, sediment, nutrients, and associated constituents to the tidal region of the Chesapeake Bay watershed (Donigian and Patwardhan 1992). Detailed applications involve extensive data requirements. It is recommended that HSPF be used only by experienced modelers.

The next 3 models discussed in Table 3-4 incorporate a relatively sophisticated representation of sediment transport mechanisms. The Channel Transport (CHNTRN) model is a 1-D time-variable model used to simulate chemical fate in well mixed estuaries (Yeh 1981, 1982; as described in USEPA 1987). It simulates sand, silt, and clay, and incorporates the degradation and sorption processes used in EXAMS. Sediment-Radionuclide Transport Model (FETRA) is a 2-D depth averaged model that simulates sediment and chemical in estuaries and coastal areas (Onishi and Thompson 1984; as described in USEPA 1987). It includes decay, settling and resuspension, and adsorption and desorption of chemical by sediments. This model also incorporates a relatively detailed representation of sediment transport processes. The Sediment Contaminant (i.e., Radionuclide) Transport model (SERATRA) is a 2-D vertical resolution model that also simulates cohesive and noncohesive transport of sediment and chemical in rivers (Lee 1979; Onishi and Wise 1982a, 1982b). It incorporates features that are similar to FETRA.

The last model discussed in Table 3-4 is RECOVERY, a U.S. Army Corps of Engineer-supported model (Boyer et al. 1994; Ruiz et al. 2000), which is an extension of frameworks developed by Chapra (1982) and Chapra and Reckhow (1983). RECOVERY represents a well-mixed surfacewater layer underlain by a vertically stratified, horizontally well-mixed sediment column. The sediment is defined by 3 zones: a well-mixed surface zone; a deep, contaminated sediment zone; and a deep, clean sediment zone. The 2 "deep" zones can be segmented into layers with varying thicknesses, porosities, and contaminant concentrations. The model includes sorption, resuspension, diffusion, first-order decay, volatilization, settling, and burial. RECOVERY's configuration is useful for evaluating capping scenarios and evaluation of sites where contamination appears layered (e.g., sites where contamination has occurred over a long time).

Finally, there are 3 additional models that have only recently been reviewed, and while not included in the accompanying tables, they warrant consideration. The first is the WASP-based Metal Exposure and Transformation Assessment Model (META4), which combines WASP5 transport with MINEQL speciation (Medine 1995; Martin and Medine 1998). This model was developed with the support of 2 USEPA laboratories for use in the evaluation of aquatic resource management strategies for metals, including waste load allocations, evaluation of remedial actions (restoration), and evaluation of total maximum daily loadings. META4 consists of a chemical equilibrium simulation submodel, based upon a solution approach similar to MINTEQ, that is coupled with the generalized WASP. The META4 simulation program is a generalized metals transport, speciation, and kinetics model developed for application to a variety of receiving waters experiencing metals contamination, including ponds, streams, rivers, lakes, and estuaries. The META4 model uses the basic transport scheme of WASP, allowing its application to a variety of water bodies in 1-, 2-, or 3-D mode, as well as the simulation of both water column and benthic layers. The META4 (Version 4.0) model is a recent version of this system which addresses metal speciation and kinetics for the chemical reactions. The solution approach is similar to that of MINTEQA2 (Version 3.11), developed and distributed by the USEPA (1991). The META4 submodel addresses some of the limitations of the WASP4 modeling system to accurately and realistically describe metal dynamics in surface waters.

Physical and chemical processes that affect the transport of metals are taken into account in the model, including advection, dispersion, chemical reaction, adsorption, desorption, erosion, sedimentation, precipitation, and dissolution. The model addresses reaction kinetics in that, when setting up the model, reactions can be included or excluded based on whether they would occur in the allowable reaction time. This time period is determined based on the water volume and flow through the compartments. Algorithms for the simulation of crucial metal transformation processes, such as aqueous speciation, sorption and desorption, chemical precipitation and dissolution, and kinetics, are included in META4 and are thoroughly described in the user manual (Martin and Medine 1998). The most significant, recent modifications to the basic META4 model have been the addition of a double-layer adsorption model to represent the interactions of dissolved metals with the iron oxyhydroxides in the water column and the benthic region and variable pH simulation capability. These additions reduce the uncertainty in the ability of the model to reasonably represent future water quality for zinc, copper, cadmium, and other sorbed metals as general major ion chemistry and metal transformation processes are varied. Some of the advantages of WASP4/META4 are that it can represent

- 1-, 2-, or 3-D environments (streams, rivers, reservoirs, multiple benthics);

- sequential deposition or scouring of benthic bed layers, transient storage;
- constant or variable pH;
- numerous point and nonpoint loads; and
- multiple metal and major ion reactions, including individual aqueous species.

The model has been applied in 7 river basins to evaluate metal impacts and remediation effectiveness, including simulations for the Clear Creek Superfund Site, Colorado, USA (CDNR/CDPHE 1996; Medine and Martin 2000a) and the Summitville Mine Site and the Alamosa River, Colorado, (CDNR/CDPHE 1996; Medine and Martin 2000b) for a variety of metals (iron, manganese, zinc, cadmium, copper, and lead). For the Clear Creek watershed, the modeling has been used repeatedly to evaluate diversion impacts, sediment restoration, the need for and effects of metal treatment plants, and the optimum location for such facilities. Metals evaluated as part of the modeling included manganese, iron, aluminum, zinc, copper, and cadmium. The modeling of the Alamosa River basin included the area upstream of Terrace Reservoir and downstream of Wightman Fork located in southern Colorado. Evaluation of current and historical water quality data was performed to provide an estimate of seasonal conditions in Wightman Fork prior to the start of mining, and these estimates were then used as input to the META4 model, simulating conditions in the Alamosa River from the confluence downstream to Terrace Reservoir.

A second model that has not been reviewed in detail herein, but one that has been applied in a fate and transport analysis of metals, is a model that was used to evaluate copper in San Francisco Bay, California, USA. Chen (1996) applied this model in a preliminary mass balance analysis of the bay. Developed from an earlier link-node hydrodynamic model of the system, the existing model was modified to include transport and fate of suspended sediment and copper. The model incorporated constant settling and resuspension rates for solids and simple equilibrium partitioning. Copper–DOC interactions were not explicitly evaluated, but rather, copper–DOC complexes were included as total dissolved copper. Fluxes between the water column and sediment were limited to settling and resuspension of particles. The model was calibrated to summer and winter seasonal average total dissolved solids (TDS; a conservative tracer), suspended solids, and copper data. The model was not identified by name and it is not known if it is available for general distribution. The authors stated that further model development was required, including the dynamic simulation of deposition and resuspension rates. Simulation of average conditions was also acknowledged as a weakness in the model. Although the description of the sediment model was not very detailed, it appears that it was relatively simple because the flux of copper between the sediment and water column was controlled only by settling and resuspension mechanisms.

Finally, a third model that was not considered in detail at the time of the original review, but that warrants recognition, is MIKE21. MIKE21 is a 3-D model that numerically solves the controlling hydrodynamic and fate and transport equations for a range of water quality variables. It is a commercially available model that may be obtained from the Danish Hydraulic Institute (DHI). Pursuant to recent releases from DHI, one of the MIKE21 modules is used to simulate the fate, transport, and bioaccumulation of metals. It represents adsorption–desorption of metals to particulate matter, settling and resuspension of sorbed metal between the water column and bed, and diffusive exchange of metal between the dissolved forms in the water column and sediment interstitial water.

Probabilistic overlays

Although probabilistic overlays are not covered in detail in this review, it is important to recognize the utility of applying a probabilistic overlay when conducting a fate and transport analysis. This is particularly true in a regulatory setting such as exists in the U.S. The USEPA has developed WQC that specify a concentration magnitude, a duration, and an allowable frequency of exceedance. The allowable frequency of exceedance is once every 3 years. To compare the modeling results to the WQC then, the predicted exposure levels need to be analyzed to determine the concentration associated with a 1 in 3-year return period.

When a steady-state analysis is performed, it is not possible to evaluate a return period for exceedances. In this case, when a steady-state model is used, an alternative is to perform a Monte Carlo analysis to synthesize a large number of sets of model inputs for the steady-state model. The results of numerous steady-state model runs may then be analyzed statistically to characterize the frequency of exceedances of the applicable WQC.

Time-variable simulations may be used in an analogous manner. That is, the output from a long-term model simulation (e.g., 20 years) can be statistically analyzed to evaluate the frequency of exceedance of the WQC. A variation of this approach is to use a Monte Carlo simulator to generate the time series of inputs (e.g., boundary conditions, loads, and flows) to the time-variable model.

Numerous statistical models are available for use in conducting these types of analyses. The Probabilistic Dilution Model (PDM) (Di Toro 1984) is used to predict the statistics of concentrations downstream from a discharge from a statistical characterization of the upstream and discharge water quality and flow. Program Monte (HydroQual 1997), developed for the Silver Council, is a Monte Carlo program that uses a characterization of the cumulative density function of concentrations and flows as inputs, as well as information on autocorrelation and cross-correlation of the variables, to synthesize a time series of daily concentrations that preserves the appropriate autocorrelation of each variable. The Dynamic Toxics (DYNTOX) model (LTI 1994), is a USEPA model that provides 3 probabilistic techniques (continuous simulation, Monte Carlo simulation, and lognormal analysis) to predict the return period of water quality standard violations. Of the fate and transport models reviewed in detail herein, only RIVRISK (EPRI 1996), a steady-state analytical solution model, and SYVAC (Bird et al. 1993), one of the time-variable analytical solution lake models, have integrated Monte Carlo routines.

Comparison of fate and transport model features

This section presents a review of the models described previously, exclusive of WQAM (a composite of numerous simple calculation procedures), but from a different perspective. The discussion is organized to highlight, across all 21 models, differences in key model features. The important features are summarized in Table 3-5, a composite of Tables 3-2 to 3-4 that is provided to facilitate a comparison of all models. The model features will be discussed in the approximate order that the information is presented in Table 3-5. The general topics to be considered are

1) fluid transport regimes and applicable water body types;
2) time and space scales and model dimensionality;
3) particulate transport mechanisms;
4) chemical partitioning and transfers;
5) degradation and transformation processes; and
6) availability of models, availability of technical support, and level of experience required to use the models.

The advantages and disadvantages of alternative model formulations will also be discussed.

Model applicability: Transport regimes and water body types

The fluid transport regime to which a model is to be applied is an important model selection criterion. Figure 3-1 indicates 3 general fluid transport regimes that are representative of most settings. Under each fluid transport regime, the models in Table 3-5 that could be applied to the particular flow regime are listed. The water flow in free-flowing streams and rivers is for the most part driven by gravity. This setting is characterized by a unidirectional, logarithmic velocity profile. All of the models that have been reviewed herein, except QWASI and USES, are applicable in this type of setting and are listed in the first column of Figure 3-1.

The second fluid transport regime considered on Figure 3-1 is a completely mixed lake. This is perhaps an idealized representation of conditions in most lakes, but it has been found to be a useful conceptual picture for screening-level analyses in many settings. As shown, SLSA, DJOC, QWASI, and USES are the analytical solution models that are applicable to this situation. Similarly, of the steady-state numerical models, CTAP and MEXAMS could also be applied. The time-variable numerical models typically provide excellent flexibility and would be applicable in

Table 3-5 Summary of features of fate and transport models

	Analytical solution model name								Numerical solution model name														
									Steady-state								Time-variable						
Characteristics and processes	WQAM	RIVRISK	SLSA	GMII	MICHRIV	DJOC	QWASI	USES[a]	CTAP	PAWTOXIC	SIMPTOX3	MEXAMS	EXAMSII	RIVEQLII	WASTOX	RCATOX	WASP5	DELFT3D	HSPF	CHNTRN	FETRA	SERATRA RECOVERY
Water body																						
River or stream	X	X		X	X			X		X	X	X	X		X	X	X	X	X	X	X	X
Lake or reservoir	X		X	X			X	X	X		X	X	X	X	X	X	X	X	X	X	X	X
Estuary	X					X	X		X						X	X	X			X		
Fluid transport																						
User specified or defined	X	X	X	X	X		X	X	X	X	X	X	X	X	X	X	X	X	X	X	X	X
Hydrodynamic submodel						X											X	X				
Time dependence																						
Steady-state	X	X	X	X	X	X		X	X	X	X	X	X	X	X	X	X	X	X	X	X	X
Dynamic			X			X	X										X	X	X	X	X	X
Dimensionality																						
1D	X	X	CM[b]	X	X	X CM	CM	X	X	X	X	X	X	X	X	X	X	X	X	X	X	X
2D (2H or 2V limitation)	2H[c]																			2H	2V[d]	
3D																	X					
Bed layers	1	1	1	1	1	1	1	M[e]	1	1	1	M	1	1	M	M	2	1	1	?[f]	M	M
Particulate transport																						
Net settling (S) or resuspension (R)									S or R		S											
Settling	X	X	X	X	X	X	X	X		X	X	X	X	X	X	X	X	X	X	X	X	X
Resuspension	X	X	X	X	X	X	X	X		X	X	X	X		X	X	X	X	X	X	X	X
Bulk exchange								X					X	X			X	X				
Bioturbation								X		X			X									
Scour or burial	X		X		X		X							X	X	X	X	X	X	X	X	X
Expanding bed or storage														C[g]	C	C	C	C	C	C	C	C
Particle types	1	1	1	1	1	1	1	5	1	1	1	M	1	1	2	3	3	3	3	3	3	1

[a]EUSES current version (EC 1996), features may differ; [b]CM = completely mixing volume; [c]2H = 2-D horizontal plane; [d]2V = 2-D vertical plane; [e]M = multiple layers; [f]? = unknown; [g]C = calculated by program; [h]B = bulk exchange (particulate and dissolved chemical); [i]V4 = SMPTOX4 only.

Table 3-5, cont'd

	Analytical solution model name								Steady-state						Time-variable									
Characteristics and processes	WQAM	RIVRISK	SLSA	GMIII	MICHRIV	DJOC	QWASI	USES[a]	CTAP	PAWTOXIC	SIMPTOX3	MEXAMS	EXAMSII	RIVEQLII	WASTOX	RCATOX	WASP5	DELFT3D	HSPF	CHNTRN	FETRA	SERATRA	RECOVERY	
Chemical reactions and transfer																								
Equilibrium (E) or nonequilibrium (N) sorption to particles	E	E	E	E	E	E	E	E	E	E	E	E	E		E	E	E	E	E,N	E	E,N	E	E	
DOC complexation											B[h]	X	B											
Water-bed diffusive exchange	X	X	X	X	X	X	X	X	X		X				X	X	X	X	X	X	X	X	X	
Inorganic speciation or complexation	X										V4[i]	X	X	X										
Precipitation or dissolution	X											X	X	X										
SEM/AVS																								
Degradation or transformation process																								
First-order decay	X	X	X	X	X	X	X	X	X	X	X		X		X	X	X	X	X	X	X	X	X	
Photolysis	X	X									X		X		X	X	X		X	X	X	X		
Hydrolysis	X	X											X		X	X	X		X	X	X	X		
Oxidation	X	X											X		X	X	X		X		X	X		
Volatilization	X	X				X		X			X		X		X	X	X	X	X		X	X		
Daughter product																X				X				
Availability																								
Public domain	X		X	X	X	X	X	X	X	X	X	X	X	?	X	X	X		X	X	X	X	X	
Proprietary		X																X						
Technical support																								
USEPA	X		X	X	X	X	?	?	X	X	X	X	X	?	X	X	X	X	X	?	X	?	X	
Non USEPA	X	X	X		X	X	?	?	X	X	X	X	X		X	X	X			?	?	?	X	

[a]EUSES current version (EC 1996), features may differ; [b]CM = completely mixing volume; [2H] = 2-D horizontal plane; [d]2V = 2-D vertical plane; [e]M = multiple layers; [f]? = unknown; [g]C = calculated by program; [h]B = bulk exchange (particulate and dissolved chemical); [i]V4 = SMPTOX4 only.

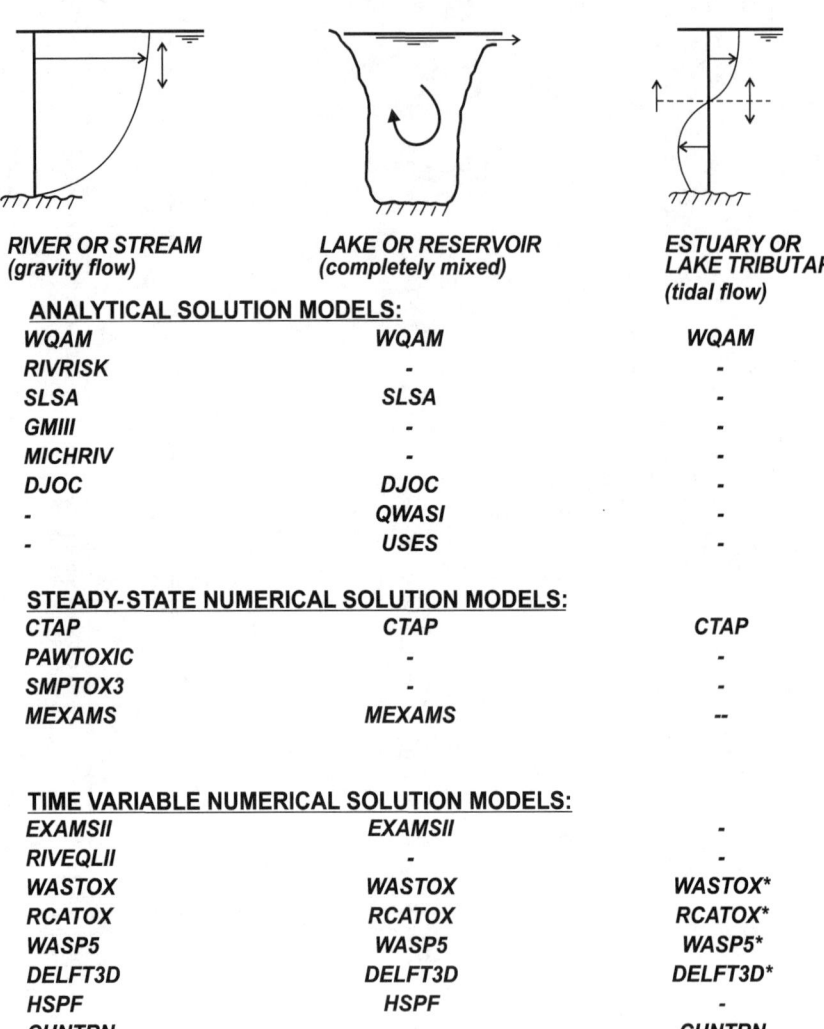

FLUID TRANSPORT REGIME:

RIVER OR STREAM (gravity flow)	LAKE OR RESERVOIR (completely mixed)	ESTUARY OR LAKE TRIBUTARY (tidal flow)
ANALYTICAL SOLUTION MODELS:		
WQAM	WQAM	WQAM
RIVRISK	-	-
SLSA	SLSA	-
GMIII	-	-
MICHRIV	-	-
DJOC	DJOC	-
-	QWASI	-
-	USES	-
STEADY-STATE NUMERICAL SOLUTION MODELS:		
CTAP	CTAP	CTAP
PAWTOXIC	-	-
SMPTOX3	-	-
MEXAMS	MEXAMS	--
TIME VARIABLE NUMERICAL SOLUTION MODELS:		
EXAMSII	EXAMSII	-
RIVEQLII	-	-
WASTOX	WASTOX	WASTOX*
RCATOX	RCATOX	RCATOX*
WASP5	WASP5	WASP5*
DELFT3D	DELFT3D	DELFT3D*
HSPF	HSPF	-
CHNTRN	-	CHNTRN
FETRA	-	FETRA
SERATRA	-	-
-	RECOVERY	-

* Reversing tides

Figure 3-1 Fluid transport regimes

this type of setting as well. One should also consider that ease of use tends to decrease from top to bottom on this figure, so for the analysis of a completely mixed system, the analytical solution models might be a useful point of departure. Of course, many of the other factors listed in Table 3-5 must also be considered and may govern the model selection decision.

The most complex fluid transport regimes are encountered in tidal estuaries and coastal regions. In estuaries, the net flow in the deeper water may be in the landward direction, counter to the net flow toward the ocean in the surface layer. Density-induced stratified conditions may also prevail. As a result of either of these conditions, a multilayer model may be called for. Models applicable in this setting are CTAP, WASTOX, RCATOX, WASP5, CHNTRN, and FETRA. Reversing tidal currents further complicate the situation. Simulation of tidally averaged conditions is one way to address this factor. If it is necessary to include simulation of tidal currents, however, models that are designed to interface with a hydrodynamic model, such as WASTOX, RCATOX, WASP5 or DELFT3D, will be required.

Time and space scales and model dimensionality

The dimensionality of the model refers to the number of directions in the X, Y, and Z coordinate system that the model can be subdivided in order to physically represent the cross-section of the receiving water. Figure 3-2 summarizes the alternatives. The upper panel illustrates what is referred to as a completely mixed volume (CMV), a simple conceptual representation of a water body that is sometimes applied to lakes. The remaining 4 diagrams illustrate the type of segmentation that would be applicable if the discharge is well mixed across the channel (1-D), vertically mixed but not laterally or horizontally mixed across the channel (2-DH), laterally mixed but not well mixed vertically (2-DV), or neither laterally or vertically mixed within the channel (3-D).

One of the 3 lower model grids illustrated on Figure 3-2 would need to be applied if accurate predictions of concentrations are required within the near field region of a discharge, prior to complete mixing across a channel. Figure 3-3 illustrates 2 spatial scales to consider when conducting a near field evaluation. First, in the immediate vicinity of the point of discharge from a pipe or a diffuser, the initial dilution that the effluent undergoes is controlled by the momentum and buoyancy of the effluent plume (Figure 3-3, left side). The spatial extent of this region is typically limited to a distance that is on the order of 10 to 100 feet from the point of discharge, and the residence time within this region is on the order of a few seconds to a few minutes. The fate and transport models reviewed herein are not applicable to this region. Rather, specialized models such as PLUMES (Baumgartner et al. 1994) and CORMIX (Jirka et al. 1996) are available to analyze this region and to predict initial dilution characteristics (Mullenhoff et al. 1985).

As stand-alone models, the plume models do not possess the capabilities included in the fate and transport models described herein (e.g., particulate transport or

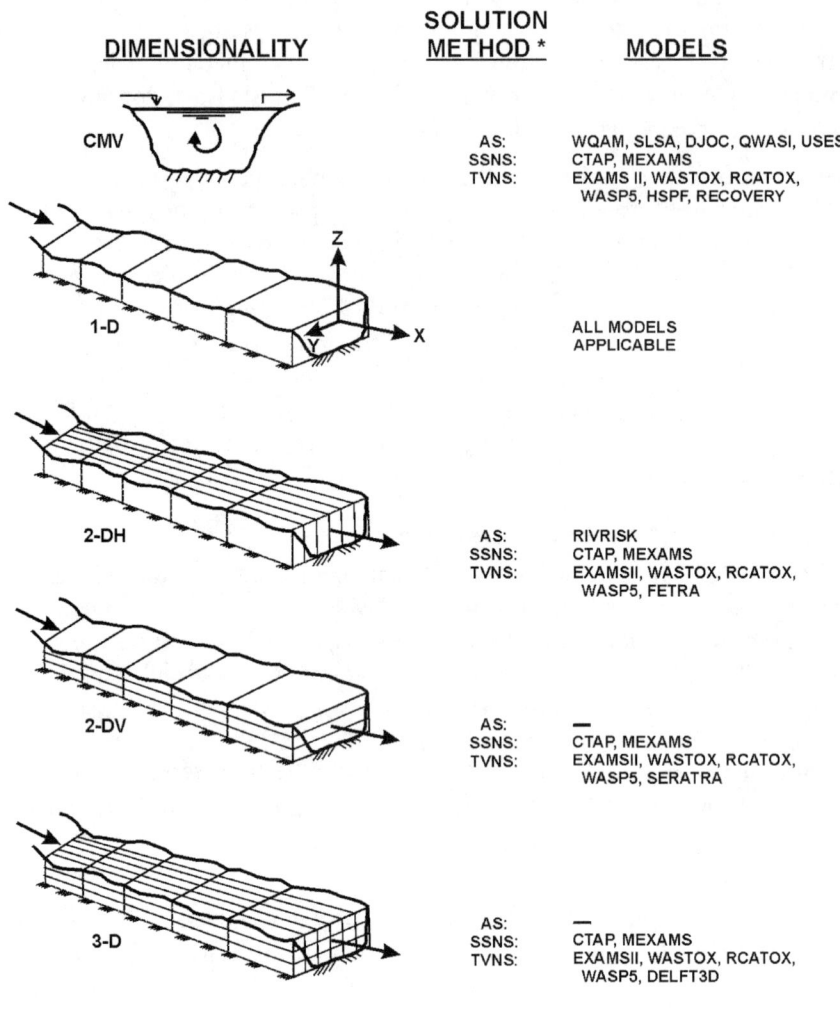

Figure 3-2 Model dimensionality

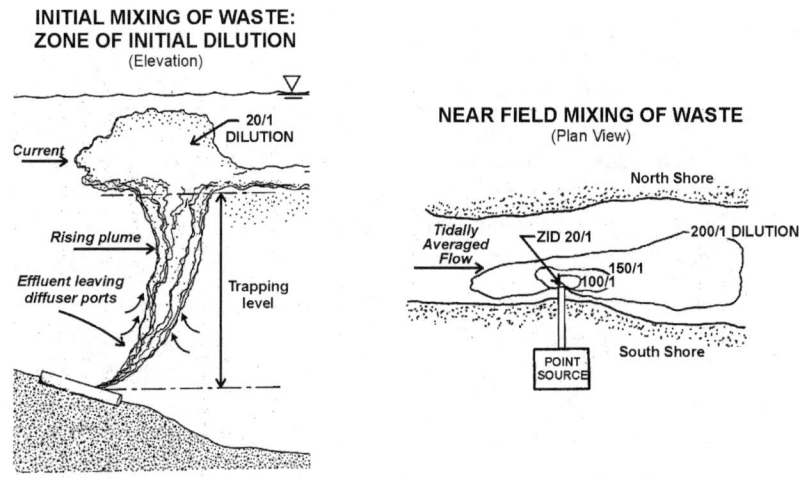

Figure 3-3 Problem settling for zone of initial dilution (ZID) and near field mixing zone analyses

partitioning), but are simply designed to evaluate the initial dilution that is achieved in the immediate vicinity of the outfall. The results of these models are often used to establish the regulatory limits of a zone of initial dilution (ZID) and to evaluate the size of an initial mixing zone for a discharge. Although the assumption of equilibrium conditions may be reasonable as a first approximation in many modeling analyses, the kinetics of complexation of metals by organic matter may be an important consideration in this region. Due to the short residence time in the ZID, this assumption should be made with caution when this spatial scale is evaluated.

The second spatial scale of interest is the area beyond the limits of the plume, but prior to complete mixing (Figure 3-3, right side), where concentration gradients are controlled by turbulent mixing processes rather than by the momentum and buoyancy of the discharge. The size of the near field region can vary from 10 to 100 feet in a small stream, to several miles or more, depending on the characteristics of the discharge and the size and shape of the receiving water. It is in this near field mixing zone and beyond (the far field) that the fate and transport models discussed herein are most appropriate for use. A 1-D model would not be applicable in this region if accurate estimates of exposure levels are desired prior to complete mixing. Rather, the multidimensional grids illustrated on Figure 3-2 need to be applied in this intermediate transition region, prior to complete mixing. Shi and Allen (1995) have shown that redistribution of metal between soluble and particulate components typically occurs within the time scale of the mixing zone. Thus, although equilibrium assumptions are more likely to be applicable in this region than within the ZID, kinetic considerations should not be overlooked.

Particulate transport mechanisms

As stated previously, particulate transport is important in that sorbed metal will be transported along with the particulate material, so the ultimate fate of a metal is dependent on having a realistic representation of particle transfers in the water body. Figure 3-4 compares the particulate transport mechanisms included in the various models that are reviewed. Each diagram schematically illustrates a water column–sediment system, with arrows indicating the direction of transfers of particulate material. The diagrams vary from the simplest representation, shown at the top of Figure 3-4, to the most complex representation, shown at the bottom.

Figure 3-4a illustrates a relatively simple particulate transport representation, with only net settling or net resuspension considered. RIVRISK employs only net settling, while PAWTOXIC can specify either net settling or net resuspension. Because it is a steady-state model, however, only one or the other of these transport mechanisms can be applied for a given simulation. With this type of model formulation, the model may be able to be calibrated for a short-term situation where one or the other conditions persist. However, in many rivers, the conditions change from net settling to net resuspension over a range of flow conditions. Thus, in a longer-term steady-state simulation, these models could only simulate the long-term average condition. If over the long-term, net sedimentation is occurring, but periodically contaminated sediments are resuspended, this type of model formulation would not properly account for a source of metals from the bed to the water column. If long-term net scouring of the bed is the prevailing condition, RIVRISK would not be applicable.

Although RIVEQLII is also listed along with this first schematic diagram of particle transport mechanisms, it is actually formulated somewhat differently from this or any of the other diagrams shown here. Based on the description in Chapman (1982) it does not appear that RIVEQLII includes particulate transport, other than settling to the bed of a metal precipitate. When suitable conditions occur, the settled precipitate undergoes nonequilibrium redissolution into the water column pursuant to a simple kinetics-based rate-limiting formulation. RIVRISK also incorporates an exchange with the bed via precipitation of metal to the bed and subsequent redissolution of metal from the bed. These processes are described in further detail in the next section.

Figure 3-4b illustrates the particulate transport mechanism included in EXAMSII. This conceptualization of water–bed interactions assumes that a bulk exchange of material (water + particles) takes place, with both dissolved metal and sorbed metal exchanged together. Because EXAMSII also allows specification of bulk exchange between bed layers (the curved arrow), a method is available to represent bioturbation, the mixing of the sediment by sediment-dwelling organisms. EXAMSII does not account for net scour or resuspension, nor does it account for changes in water column or bed solids concentrations over time. A disadvantage of only using bulk exchange with the water column is that the model is not sensitive

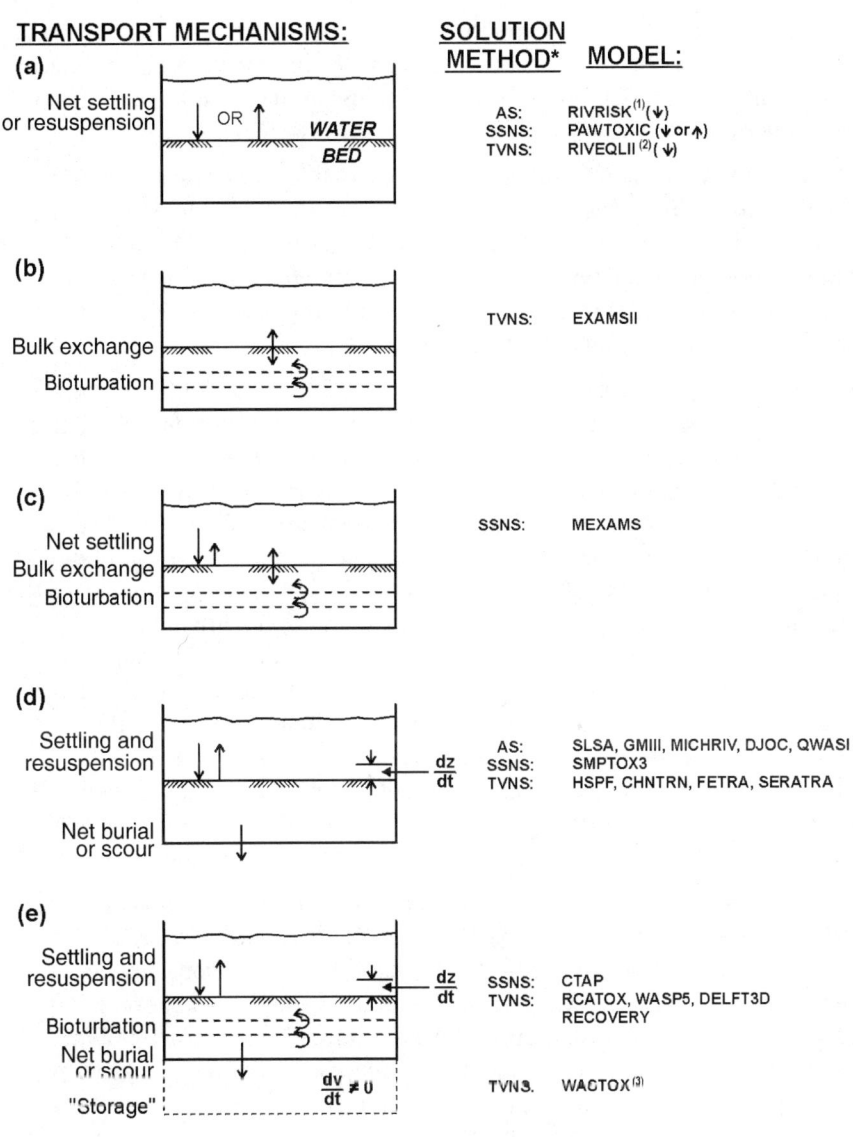

Figure 3-4 A comparison of alternative particulate transport representations

Notes: USES - Equilibrium partitioning between water column and bed
(1) Precipitation to fixed surface and redissolution to water column.
(2) Settling of precipitated metal with rate limited redissolution.
(3) includes DOC complexation.
(4) With storage or expanding bed option.

to relative differences in the controlling water column–bed transfer processes that apply to strongly sorbed substances, where particle transport controls, and weakly sorbed compounds, where the diffusive flux of dissolved chemical is a relatively more important process.

The MEXAMS version of EXAMS provides a somewhat refined representation of particle transport by including settling and resuspension of solids (Figure 3-4c), constrained by the requirement that the net solids balance yields a condition of net sedimentation. Although this provides an improved representation of sediment–water column exchange processes, diffusive transport of dissolved chemical is still not explicitly accounted for with this approach.

The fourth particle transport schematic (Figure 3-4d) includes both particle settling and resuspension and deep burial of solids (the downward-pointing arrow beneath the sediment layer; computed by the model from the difference between settling and resuspension fluxes). This representation is used by SLSA, GMIII, MICHRIV, DJOC, and QWASI (all analytical solution models); by SMPTOX3, a steady-state numerical model; and by HSPF, a time-variable numerical solution model. It is understood that CHNTRN, FETRA, and SERATRA also conform to this configuration. Particle mixing between bed layers due to bioturbation is not included in these models because the sediment consists of a single completely mixed layer. The notation dz/dt represents the change in elevation in the vertical or z coordinate direction of the surface of the bed over time. It represents the net rate of sedimentation or scour over time and is computed by the model from the difference in settling and resuspension rates of particles at a fixed bed solids concentration. Net sedimentation is an important fate-controlling process for contaminated sediments and is often used as an aid in model calibration.

Finally, Figure 3-4e includes the same mechanisms as in the preceding case, plus mixing between bed layers due to bioturbation. This framework is used by the steady-state model CTAP and by the time-variable numerical models WASP5, WASTOX, RCATOX, and RECOVERY. A "variable volume" or storage option is also included in WASTOX. This option provides an alternative way to account for burial of sediment chemical. Without this option, the model does not keep track of chemical buried beneath the deepest model layer. Hence, when a period of net scour occurs, there is not any mechanism to reintroduce this material into the system. The variable volume option stores the buried chemical in a completely mixed bottom layer that expands or contracts as net sedimentation or scour occurs.

Chemical partitioning and transfers

Chemical partitioning—An important reason why many of the available fate and transport models are not well suited for use with metals is that the partitioning of metals between the dissolved and particulate phases is not well represented. The accurate evaluation of this distribution of the metal is important because it has a direct effect on the magnitudes of the dissolved and particulate transport mechanisms that control transfers of the metal within the system. This evaluation is also important because it has a direct bearing on the bioavailability of the constituent of interest. A brief description of the alternative formulations commonly used in most fate and transport models will follow. The underlying equations for 2-phase and 3-phase partitioning are derived and discussed more formally by O'Connor (1988a). A detailed discussion of alternative approaches to chemical speciation and complexation modeling is reserved for Chapter 4, where chemical equilibrium models are reviewed.

Figure 3-5 illustrates some of the ways partitioning is represented in the models reviewed herein. The simplest representation, 2-phase equilibrium partitioning, is illustrated on Figure 3-5a. Here, equilibrium partitioning is assumed to occur between the dissolved and particulate phases in both the water column and bed sediment. Sorption to particles is linear with respect to concentration, with the bound chemical concentration proportional to the dissolved concentration in the water. Only total dissolved chemical is considered, even though it is known that the total dissolved chemical includes both dissolved inorganic species and DOC-complexed chemical. The dissolved fraction is typically taken to be the bioavailable fraction of chemical. The distribution of chemical between the dissolved and particulate phases is evaluated from the suspended solids or particulate organic carbon (POC) concentration and the partition coefficient for the chemical of interest.

With the appropriate specification of inputs, all of the models reviewed herein can be used with the 2-phase formulation of partitioning. The TGD (Appendix VIII of EC 1996a) recommends use of measured partition coefficients for metals when this approach is used. The TGD also suggests that the partition coefficient be based on the "available" concentration, rather than the total concentration in the solid and liquid phases, in recognition of the fact that metal in the mineral fraction is not available. The exact procedure for these types of evaluations is not well understood and is an area of ongoing research. The TGD also highlights the need for both the PEC and PNEC to be estimated on the basis of "similar levels of availability." This reasoning is consistent with the USEPA's approach to apply conversion factors to convert the WQC to a dissolved metal basis and to use a metal translator in the receiving water to account for the effects of partitioning to suspended matter (USEPA et al. 1996).

The next level of refinement in fate and transport models is to employ 3-phase equilibrium partitioning, as shown schematically on Figure 3-5b. Here, the freely dissolved chemical is in equilibrium with the chemical sorbed to particles and the chemical complexed to DOC. In this case, the DOC is operationally defined as the fraction of the total organic carbon (TOC) that passes through a 0.45 μm filter. It therefore includes colloidal organic matter. The DOC-complexed chemical is represented in the model in much the same way as the sorbed chemical in the 2-phase partitioning model. In this case, the total dissolved chemical includes the

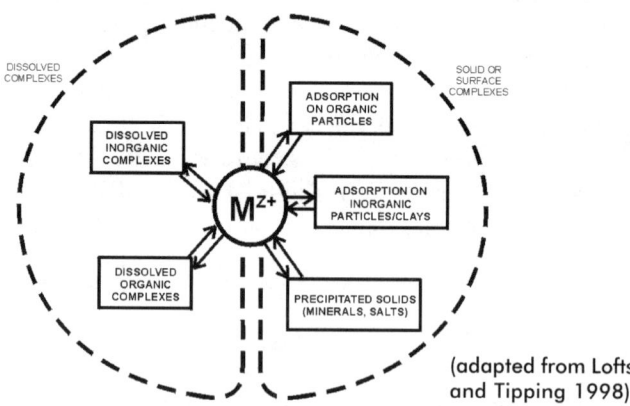

Figure 3-5 Alternative partitioning approaches

DOC-complexed chemical plus the freely dissolved or uncomplexed chemical. Very often in modeling analyses only the freely dissolved chemical is considered to be bioavailable. The USEPA WQC, such as the WQC for copper as of 2003, are compared to this total dissolved metal concentration, even though the DOC-complexed fraction is relatively nonbioavailable. Of the models included in Table 3-5, a 3-phase equilibrium partitioning formulation is included in EXAMSII, WASTOX, RCATOX, WASP5, and DELFT3D. DELFT3D simulates DOC complexation by setting the sedimentation rate to zero in one of the 5 organic carbon pools that are included in the model (L. Postma, Delft Hydraulics, personal communication 1999). It is noted that EXAMSII is unique in that it also includes sorption to biological particles as a distinct sorption phase.

The most refined approach to evaluation of partitioning in a fate and transport model is to incorporate a chemical equilibrium model in the analysis. Figure 3-5c, described in more detail in Chapter 4, illustrates the important components that would be incorporated in a full-featured implementation of this approach. Of the models reviewed herein, only RIVRISK (using a selection of predetermined speciation calculations), MEXAMS (using MINTEQ), RIVEQLII (using MINEQL), and DELFT3D make use of this general approach. Of these 4 models, the approach used by the latter three is preferred because speciation is computed directly from the water quality characteristics at the site of interest. The most recent version of DELFT3D incorporates the nonideal competitive adsorption (NICA) formulation for metal–DOC complexation (Koopal et al. 1994; Benedetti et al. 1995; Kinniburgh et al. 1996; Temminghoff et al. 1997; de Rooij et al. 1999). The NICA formulation is a multisite model that is comparable in structure to the Windermere Humic Aqueous Model (WHAM) (Tipping 1994) that is described in detail in Chapter 4.

Although each of the models in this review has strengths and weaknesses, with regard to speciation, those with an internally linked metal speciation routine are considered more sophisticated than the others. However, the other models can make use of stand-alone chemical equilibrium models to provide a more detailed characterization of metal speciation and partitioning. Several general guidance documents that discuss methods to evaluate partition coefficients are noted in the summary of guidance documents that concludes Chapter 3 and a detailed review of the important chemical equilibrium models is presented in Chapter 4. Given the increased awareness of the importance of evaluating metal speciation in assessing bioavailability, it is envisioned that geochemical models will be used with increasing frequency in the future when fate and transport analyses for metals are performed.

Chemical transfers—The chemical transfers in a fate and transport model may be categorized as particulate and dissolved transfers. The partitioning method will define the fraction of chemical associated with each of these phases. The dissolved phase will include the freely dissolved chemical and that which exists as dissolved organic and inorganic complexes. The particle transfers were discussed previously, with Figure 3-4. Any chemical bound to the particles will be transferred with the particles, so this same figure serves to illustrate the routes of particulate chemical transfer in the various models as well. Figure 3-6 is similar to Figure 3-4, except that the dissolved chemical transfers have been added as well. Here, the letters P and D are indicated, to distinguish between particulate and dissolved chemical transfers.

Figure 3-6a does not indicate dissolved chemical transfer between the water column and the bed because these first 3 models do not include a diffusive flux of dissolved chemical between the pore water and the overlying water column. As noted previously, it appears that RIVEQLII includes only settling of a metal precipitate to the bed and not metal sorbed to particulate matter (Chapman 1982). This settled precipitate subsequently undergoes nonequilibrium redissolution into the water column in accordance with a pseudo-kinetics rate-limiting formulation. RIVRISK (EPRI 1996), in addition to net settling of metal sorbed to particulate matter, also includes an exchange with the bed via precipitation onto a fixed surface and subsequent redissolution of metal from the bed to the water column. The fact that the precipitation is to a fixed surface suggests that this precipitated metal is not subject to transport within the system.

EXAMS II transfers dissolved and sorbed chemical between the water column and the bed via a bulk exchange of chemical (Figure 3-6b). As noted previously, this could potentially complicate the representation of the water column–bed transfer rates of chemicals having markedly different partitioning characteristics. While MEXAMS (Figure 3-6c) uses this same approach, the introduction of net settling gives it an enhanced capability to represent sediment–water column transfers of chemical relative to EXAMSII. The fact that MEXAMS incorporates a refined chemical equilibrium model is another distinct advantage of this model. It is significant to note that, as a general observation, the 3 models that incorporate some form of chemical equilibrium computation in the fate and transport model (RIVEQL, RIVRISK, and MEXAMS) also tend to utilize relatively simple representations of particulate and dissolved chemical transfers. Hence, there are clearly tradeoffs that will need to be considered when selecting a model for use.

In contrast to the preceding models, some of the other models that are available for use do incorporate distinct particulate and dissolved transfer processes (Figure 3-6d and e). SLSA, a relatively simple analytical solution model, is an example of one of the models that uses this approach (Di Toro and Paquin 1984). It was tested using data from a field study in which both DDE and lindane, organic chemicals with markedly different partitioning and fate characteristics, were instantaneously released to the water column of a quarry, and the concentrations in the water and sediment were monitored over time (Waybrant 1973). SLSA successfully reproduced the water and sediment concentrations throughout the 1-year post-release monitoring period. Perhaps more importantly, the calibrated model was also able

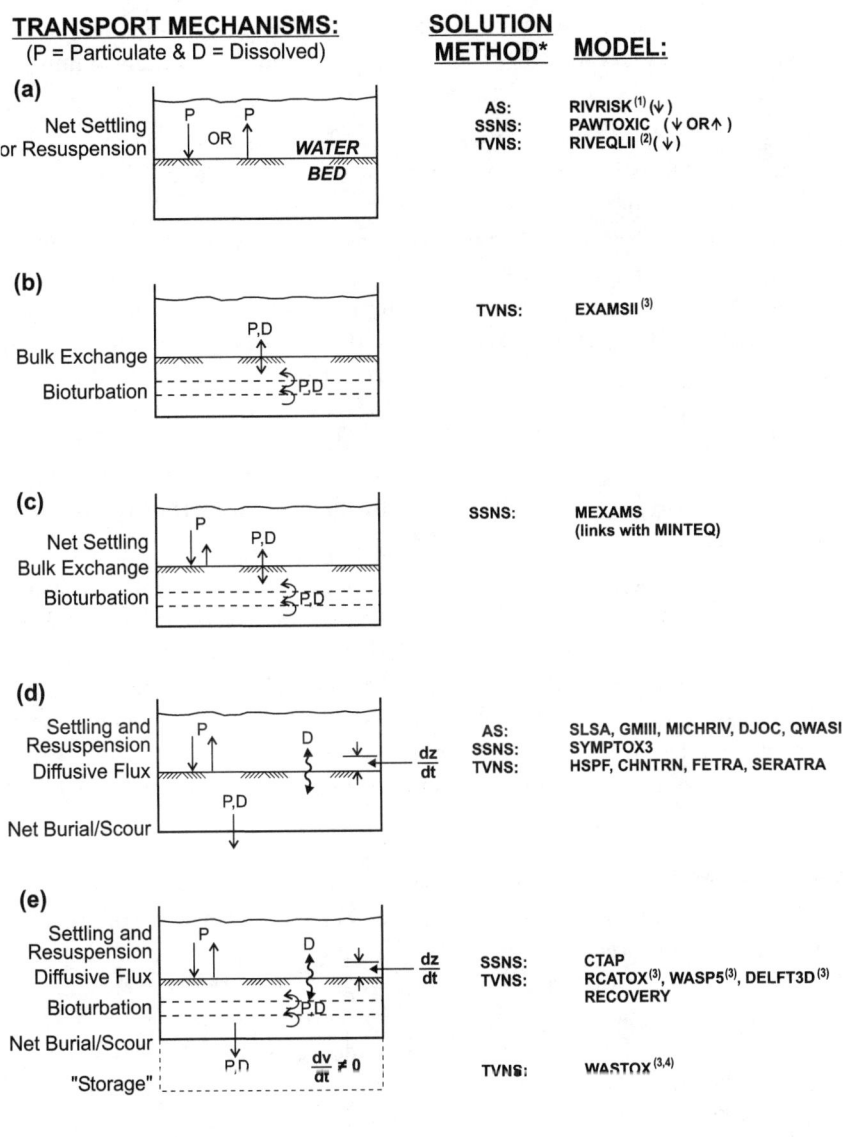

Figure 3-6 A comparison of alternative representations of chemical partitioning and water–bed interactions

Notes: USES - Equilibrium partitioning between water column and bed
(1) Precipitation to fixed surface and redissolution to water column.
(2) Settling of precipitated metal with rate limited redissolution.
(3) includes DOC complexation.
(4) With storage or expanding bed option.

to successfully predict the measured concentrations of each of these compounds in the quarry nearly 5 years after they were initially released to the water column (Di Toro and Paquin 1984).

The remaining models listed in Figures 3-6d and 3-6e also include a diffusive flux of chemical from the sediment pore water to the overlying water column. A proper evaluation of partitioning is important in this type of model if the dissolved chemical concentrations and diffusive fluxes are to be properly simulated. The partition coefficient in models that incorporate 2-phase partitioning should be based on the total dissolved chemical, while for the 3-phase models, it should be based on non-DOC complexed chemical. The latter approach is consistent with the TGD recommendation that the partition coefficient should be based on the bioavailable chemical concentration (EC 1996a). The fact that a significant fraction of the dissolved chemical may be complexed to DOC will serve to reduce the bioavailability of the total dissolved chemical concentration. The discussion of the Biotic Ligand Model (BLM) in Chapter 5 will highlight the importance of this effect and why it occurs.

It should also be noted here that although none of the models in this review include the simulation of SEM and AVS, a version of SMPTOX has been applied that does so. The model considers copper, cadmium, nickel, lead, and zinc (Dilks et al. 1995). It is a steady-state model, however, so it does not have the capability to simulate the observed seasonal variation in AVS. Additionally, it does not address the potential oxidation of AVS in sediments, a process that may result in the release of metal from the sediment to the overlying water column. These issues will be discussed further in Chapters 5 and 6.

Degradation and transformation processes
Most of the fate and transport models in Table 3-5 include some form of first-order decay process. The procedure for evaluating the rate coefficients and the degree of detail needed to specify that model inputs vary significantly, from models that accept a relatively simple lumped first-order decay rate to EXAMS, which has a detailed set of options for incorporating independent rates of photolysis, hydrolysis, biolysis, oxidation, and volatilization. It also calculates the production of daughter products. Although this level of complexity may be useful in the case of some organic chemicals, it is of lesser utility for metals.

Availability of models, documentation, and technical support
The availability of models, documentation, and technical support at the time of this review is indicated in the lower section of Table 3-5. The appendix to this book contains a partial listing of sources for models reviewed in this report.

Summary of Model Reviews and Guidance Documents

A number of useful technical references and reviews of models were identified in the course of the literature search. These documents provide sources of information for an individual working with the types of models described in this review. They include descriptions of some of the more widely used models and important model features and provide information on sources of documentation and availability of technical support. The main limitation of these references was that the focus was generally not on metals or on the applicability of the models for use in metal fate and transport evaluations. Also, although much of the general technical information on the conceptual basis of fate and transport models and alternative analytical approaches is useful, some of the information related to specific models that is included in the earlier documents is outdated.

Mills et al. (1982a, 1985), in several of the earlier publications, describe screening-level analyses that can be applied in water quality assessments for toxic and conventional pollutants. These documents include a collection of formulas, tables, and graphs for use in preliminary assessments, and computations can be performed with a handheld calculator. Although not specifically directed at evaluations for metals (the latter publication does devote a section to metals), these guidance documents provide a series of relatively simple, model-oriented procedures that can be used in a preliminary assessment of metal fate and transport. Lakes, rivers and streams, impoundments, and estuarine settings are discussed in both documents. The earlier publication has been cited in a review of models for use in exposure assessments (USEPA 1987), where it is referred to as the "WQAM." The 1985 publication includes procedures for analyzing groundwater settings as well. Documents such as these are particularly well suited for use by an analyst who is relatively inexperienced in modeling.

Delos et al. (1984) also provide general guidance for performing waste load allocations and review alternative models that can be used in this regard. The key features are summarized in tabular form and advantages and disadvantages of selected models highlighted. MICHRIV is described in detail.

Schnoor et al. (1987) provide a detailed description of fate and transport model process formulations and a compilation of model input parameters. Detailed descriptions of 3 fate and transport models (EXAMS, TOXIWASP, HSPF) and a geochemical model (MINTEQ) are included. Guidance on model selection is also provided. Although this reference is not strictly applicable to metals, it does include a section on reactions for metals. This document has not been updated but does serve as a useful resource for modeling purposes. It includes a case study waste-load allocation for zinc and copper using a simple 1-D model of the Deep River, North Carolina, USA. MINTEQ is used as a stand-alone model to evaluate metal speciation at different points along the river. A similar type of analysis, performed with an updated geochemical model, would be suitable for screening-level purposes.

The USEPA (1987) published a review of surfacewater quality models and selection criteria for models used in exposure assessments. The review discusses nonpoint source runoff models, hydrodynamic models, and contaminant transport models. It includes descriptions of some of the models discussed herein (WQAM, SLSA, MICHRIV, CTAP, EXAMS, MEXAMS, TOXIWASP, WASTOX, HSPF, CHNTRN, FETRA, and SERATRA). In a similar, more recent review, the USEPA has compiled an updated review of modeling tools for use in developing TMDLs for watersheds (USEPA 1997). Hydrodynamic models, steady-state and dynamic water quality models (including USEPA screening procedures, EXAMSII, SMPTOX3, DYNTOX, WASP5, and HSPF), and mixing zone models (CORMIX and PLUME) are described. Relatively little consideration is given to metals in either document, although MEXAMS is one of the models included in the earlier document.

Finally, as discussed previously in this chapter, the partition coefficient is an important parameter that must be evaluated when applying standard fate and transport models for metals. Two recent documents have been published in support of this type of application and should prove useful to those who are working in this area. The first report presents ways to evaluate partition coefficients for use in both surfacewater models and multimedia models (USEPA 1999b). Partition coefficient data from the scientific literature are summarized for a variety of metals, and statistical methods and geochemical models are also used to estimate partition coefficients for some metals. The second report emphasizes the importance of using site-specific data, whenever possible, in support of the evaluation of partition coefficients, and also discusses the key geochemical processes that affect partitioning (USEPA 1999c).

Chapter 4
Chemical Equilibrium Models

Historical Model Development

Chemical equilibrium models are useful for describing the distribution (speciation or complexation) of substances in aquatic systems. Since the development of the first computer codes for equilibrium calculation in the 1960s, many different individual models have been developed for a wide variety of applications and with diverse capabilities (Nordstrom et al. 1979; Bassett and Melchior 1990). The majority of these models were developed to simulate a predetermined set of chemical reactions and are therefore applicable in limited settings. Due to the large number of available models, this review is not intended to be exhaustive, but will focus on those widely used models that are actively maintained and that have unique and noteworthy features. Models with historical ties to some of the current models will also be mentioned. Although an area of active research and development, nonequilibrium models are not included in this review.

Chemical equilibrium models have several critical roles in developing risk assessments for metals. Chemical equilibrium models can be used to calculate the distribution of metals among various forms, including dissolved inorganic and organic species, amounts adsorbed on particle surfaces, and amounts in mineral solid phases (Figure 4-1). This distribution should be considered in predictions of environmental fate and transport (and therefore determination of predicted environmental concentrations [PECs]) because metals associated with minerals or suspended particles will have a different fate than those dissolved in the water column. Because aqueous phase and surface phase complex formation are simultaneous and

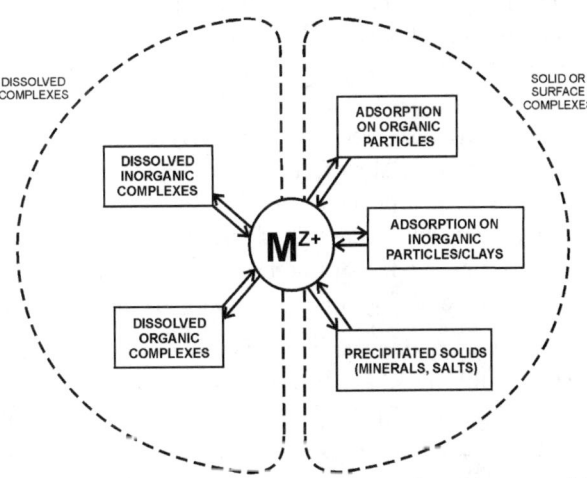

Figure 4-1 Schematic diagram of a chemical equilibrium model for metal speciation (adapted from Lofts and Tipping 1996)

competitive processes, it is somewhat of an oversimplification to represent partitioning of metals with a simple linear partitioning formulation, as was described in Chapter 3. Estimates of partition coefficients for metals tend to be highly variable due to a dependency on water chemistry parameters such as pH or the presence of metal complex-forming ligands such as dissolved organic matter (DOM) (Hudson 1998; Sigg 1998). Use of these simple approaches should therefore be limited to very narrow ranges of environmental conditions, with the partition coefficient best measured in the receiving water of interest under conditions similar to the conditions to be modeled (EC 1996a). Alternatively, a more refined, mechanistically based approach can be used that makes use of equilibrium speciation and includes aqueous and surface complexation reactions to model metal sorption (Shi et al. 1998).

A chemical equilibrium description of metal speciation can also be used to predict metal bioavailability. Free metal ion concentrations have been shown to be better predictors of metal bioavailability and toxicity than total or dissolved metal concentrations (Sunda and Guillard 1976; Campbell 1995). Free metal ion concentrations are dependent on a number of factors, including available ligands, adsorption on suspended particles, and mineral solubility (Figure 4-1) (Morel and Hering 1993). The use of chemical equilibrium models to calculate metal speciation as an estimate of metal bioavailability recently has been extended to include sensitive receptor sites on an organism as a competing ligand within the equilibrium framework (Playle et al. 1993a, 1993b; Janes and Playle 1995). The Biotic Ligand Model (BLM), discussed in Chapter 5, recently has gained increased attention from scientific and regulatory interests as a tool for determining metal bioavailability (Renner 1997; Paquin et al. 1999; Di Toro et al. 2001; Santore et al. 2001). As such, these models could be used to develop a technically defensible predicted no-effect concentration (PNEC) or water quality criterion (WQC) that reflects the effect of site-specific conditions on the bioavailable metal concentration.

Many of the existing equilibrium models have evolved from early computer codes, including RAND, REDEQL, and WATEQ (Morel and Morgan 1972; Truesdell and Jones 1973). Both REDEQL and WATEQ utilized a numerical solution scheme based on mass action expressions and therefore required a database of thermodynamic stability constants. Alternatively, the RAND program used a minimization of free energy approach for the numerical formulation.

The free energy minimization approach has been credited with several advantages including greater numerical stability and no requirement for specification of starting values. Furthermore, new reactions can be formed from existing thermodynamic information without the need to specify additional stability constants. This alternative approach combines standard free energies with molar concentrations to determine a total free energy function for the system. This function is then minimized to find an equilibrium solution. Because these models rely on minimization techniques, potential problems can arise from convergence on local rather than global solutions. Recently, the free energy minimization approach has been used in GMIN (Felmy 1995), which relies on sophisticated numerical methods to improve convergence and to avoid local minima. Despite these advantages, there are relatively few models that use this numerical scheme.

In contrast, the mass action approach requires the solution of a set of simultaneous nonlinear equations (the number of equations equal to the number of independent chemical components in the simulation). Solution techniques usually involve an iterative Newton-Raphson approach. Advantages of the mass action approach include convergence on a unique solution, greater efficiency when scaled to larger problems, and adaptability to empirical reaction formulations (such as simple K_D partitioning).

Several models were developed by the U.S. Geological Survey (USGS) from an ion association model for seawater (Garrels and Thompson 1962), including WATEQ (Truesdell and Jones 1973), SOLMNEQ (Kharaka and Barnes 1973; Perkins et al. 1990), PHREEQE (Parkhurst et al. 1980; Plummer and Parkhurst 1990), and EQ3/EQ6 (Wolery 1979; Wolery et al. 1990). Application to mineral equilibria in groundwater required that these models specialize in mineral saturation indices as well as aqueous speciation. PHREEQE contained additional functionality in its ability to trace reaction paths along phase boundaries that optionally included pH and pE changes. Development of WATEQ through several generations, including WATEQF (Plummer et al. 1976) and WATEQ3 (Ball et al. 1981), resulted in the compilation of a USGS-maintained thermodynamic database (the WATEQ lineage later included WATEQIVF; Ball et al. 1987).

A different lineage of model development beginning with REDEQL focused on establishing a compact numerical algorithm for chemical speciation. Expansion of this algorithm-based approach to larger problems led to the development of MINEQL (Westall et al. 1976), which included gaseous and solid-phase equilibrium, redox, and electrical double-layer adsorption in addition to aqueous speciation. A less comprehensive version of MINEQL called MICROQL was also developed for classroom microcomputer use (Westall 1979). A merging of the numerical algorithm from MINEQL with the thermodynamic database of WATEQ3 resulted in the development of MINTEQ (Felmy, Girvin, Jenne 1994; USEPA 1991). Further development of the MINTEQ code and database resulted in the release of MINTEQA2 along with the PRODEFA2 user interface (Allison et al. 1991). A subsequent independent pairing of the MINEQL code with a graphical user interface and the MINTEQA2 database lead to the development of MINEQL+ (Schecher and McAvoy 1992).

The Windermere Humic Aqueous Model (WHAM) was recently developed to simulate chemical equilibrium of waters, sediments, and soils dominated by natural organic matter (Tipping 1994). The WHAM was a recent addition to a series of models developed to describe natural organic matter chemistry and interactions with metals. Models I and II developed a framework for describing interactions of protons, aluminum, and calcium with humic substances (Backes and Tipping 1987). Model III added site heterogeneity (Tipping et al. 1988, 1989). Model IV added nonspecific electrostatic binding (Tipping et al. 1990, 1991; see also the Complexation by Humic Acids in Organic Soils [CHAOS] model, Tipping and Hurley 1988) and desorption of humic substances from soil horizons (Tipping and Woof 1990, 1991). Model V included trace metal complexation (Tipping 1993). WHAM Version 1.0 was based on Model V and formalized the trace metal complexation approach by establishing a database of best-fit parameters for simultaneous calibration to a wide array of metals (Tipping 1994). Recently, Model VI was developed with the addition of tridentate metal complexes and more detailed descriptions of binding heterogeneity (Tipping 1997).

In the WHAM, metal–organic matter interactions are simulated as a combination of chemical and electrostatic interactions. Competitive binding of metals includes monodentate and bidentate interactions. Chemical binding can be modified by electrostatic interactions, depending on the net charge of the organic molecules and the ionic strength of the aquatic simulation. Nonspecific electrostatic binding is also represented using a Donnan-type diffuse layer model. These capabilities, along with extensive calibration to published data sets, make the WHAM arguably the most comprehensive model to date for simulation of metal chemistry where interactions with natural organic matter are important.

Another recently developed model is Chemical Equilibrium in Soils and Solutions (CHESS) (Santore and Driscoll 1995). The CHESS model was designed to be a research and teaching aid for soil chemistry. One of the design goals was to develop a model that could be used to compare different reaction formulations for describing adsorption on natural surfaces. The CHESS model was also designed to work as a subroutine within other models to facilitate adding chemical equilibrium and speciation calculations to fate and transport and nutrient cycling models. The CHESS model has been added to terrestrial nutrient cycling models (Postek et al. 1995; Yanai et al. 1996; Yanai and Santore 1996), soil development models (Santore et al. 1995), and recently a BLM for predicting acute metal toxicity (Di Toro et al. 1997; Paquin et al. 1998; Santore et al. 1998; Meyer et al. 1999).

The CHESS model was designed around an efficient numerical algorithm for generic speciation problems. In this regard, it is similar to models such as REDEQL and MINEQL. The advantage to using a generic numerical algorithm is that new reactions can easily be added or modified to simulate increasingly complex systems. The lack of a generic computation framework in the WHAM, on the other hand, makes it difficult to add chemical processes that are currently not represented (e.g., precipitation of insoluble metal salts). Another advantage to a generic approach is that all of the necessary information for specifying the number and types of reactions is included in the input file. The current structure of the WHAM includes a description of the distribution of organic functional groups in the source code, making it difficult to change this distribution without altering the computer code. To alleviate these problems, the CHESS model was recently modified to duplicate the chemical and electrostatic interactions derived for the WHAM (Santore et al. 1998; Meyer et al. 1999). This combination provides a comprehensive description of metal–organic matter speciation within a generic speciation framework.

Finally, the Nonideal Competitive Adsorption (NICA) model (Koopal et al. 1994; Benedetti et al. 1995), a late addition to this review and not described in detail herein, is a multisite model that accounts for nonideal binding of metal ions to heterogeneous organic matter. The model was subsequently refined to include a Donnan-type formulation for nonspecific binding of ions (Kinniburgh et al. 1996). The current version is somewhat comparable in structure to the WHAM described previously. NICA has been shown to effectively simulate the binding of metal ions (e.g., copper, cadmium, and lead) to dissolved organic matter (DOM) (Koopal et al. 1994; Benedetti et al. 1995; Kinniburgh et al. 1996; Temminghoff et al. 1997; de Rooij et al. 1999).

Comparison of Models

Previous efforts to compare the output from numerous chemical equilibrium models have shown generally good agreement for major ions, but order of magnitude discrepancies in the comparison of minor species (Nordstrom et al. 1979). This is a problematic result for models applied to metals chemistry because metals species usually exist in minor or trace concentrations. However, the output from chemical equilibrium models can be verified by checking that the principles of thermodynamics and conservation of mass are consistent with the model output. It is usually trivial, therefore, to determine that a chemical equilibrium model is coded and operating properly. Given the same system and assumptions, the result should be entirely independent of the code used to generate it. Discrepancies between different models, then, can generally be attributed to differences in one of the following areas.

Major discrepancies include the following:
- List of species. Especially for models with thermodynamic databases, it is imperative that the user check the list of species included in a simulation and that all important species for a given problem are represented.
- Values for equilibrium constants (Martell and Motekaitis 1992). Furthermore, errors can be introduced to distributed thermodynamic databases where they can be difficult to detect (Serkiz et al. 1996). Even among credible laboratories, a range of values may be reported. It is imperative that values for thermodynamic constants are cross-checked against reliable sources (e.g., Smith et al. 1998), to make sure they are appropriate. Although the current trend toward user-friendly interfaces and supplied

thermodynamic databases makes it tempting to treat models as "black boxes," users of these models should, as a minimum, know how to document the reactions and equilibrium constants used in a simulation so that these can later be scrutinized along with any model results.
- Different formulations used to represent similar processes. For example, adsorption of metals to particulate material may be represented in a variety of ways (Table 4-1).

The availability of superficially similar processes in different models can be misleading if the underlying physical and chemical assumptions are different. These discrepancies tend to be greatest for chemical processes such as adsorption on surfaces or complexation with organics, where competing theories can be used to develop models with approximately the same capacity for explaining a given data set. For example, Santore and Driscoll (1995) demonstrated that different formulations for cation exchange reactions (with identical stoichiometry and thermodynamic values) resulted in marked differences in results. Similarly, Westall and Hohl (1980) showed that different electrostatic adsorption models could be calibrated to yield equivalent results, but that different values of physical and thermodynamic parameters would be required.

Minor discrepancies include the following:
- Ionic-strength effects. The use of different methods (e.g., Davies, Debye-Huckel, extended Debye-Huckel) for calculation of ion activity coefficients can lead to differences in results (Stumm and Morgan 1981). Assuming that they are not used beyond their intended ranges of ionic strength, these discrepancies should be small.
- Differences in floating point precision. The use of different levels of precision in different computer codes, or compiling the same code on machines with different floating point representations, can lead to small differences in model output.

As a result of these potential pitfalls in the application of chemical equilibrium models, there is added responsibility on the part of the user to verify the appropriateness of a given model, including specifications for all available options for an application. A comparison of features of some of the models mentioned in this review shows they have many more similarities than differences (Table 4-1). Any of these models would give nearly identical results if restricted to aqueous inorganic metal speciation (subject to the recommendations given above to determine appropriate reactions and equilibrium constants independent of any supplied thermodynamic database). Differences between models start to become more significant when the simulations involve organic matter complexation and surface adsorption processes. To some extent, these differences are due to the use of alternative formulations, based on either simple theories (e.g., partitioning) or relatively complex theories (e.g., electrical double layer) for simulating the same process (particulate adsorption). Another factor is that the natural variability in

Table 4-1 Comparison of features of selected chemical equilibrium models for metal speciation

Feature	MINEQL	MINTEQ	MINEQL+	WHAM	CHESS
Thermodynamic database supplied	No	Yes	Yes	Yes	No
User added reactions	Yes Mandatory	Yes	Yes	Yes Difficult	Yes Mandatory
Activity corrections (inorganic ions)	Davies	Davies	Davies	Extended Debye-Huckel	Davies or Extended Debye-Huckel
Temperature corrections	Yes	Yes	Yes	Yes	Yes
Speciation of natural organic matter including metal complexation	Simple discrete	Simple discrete or composite ligand model (see Allison and Perdue 1994)	Simple discrete	Discrete plus Donnan (extensive database)	Triprotic or discrete plus Donnan
Adsorption on natural surfaces	Electric double-layer (e.d.l.) model (using STANFORD Module)	Activity K_d, Langmuir, Freundlich, Ion exchange, constant capacitance, diffuse-layer, 3-layer	Langmuir, Freundlich, constant capacitance, 2-layer, 3-layer	Ion exchange in soil version only	Langmuir, Freundlich, surface complexation, 2-layer e.d.l., Ion exchange (Vanselow, Gaines-Thomas)
Solid or gaseous phase solubility	Yes	Yes	Yes	Limited to $Al(OH)_3$, $Fe(OH)_3$ in soil version only	Yes
Redox	Yes	Yes	Yes	No	No
User interface	No	Yes Text only	Yes DOS (version 3) or Win95 (version4)	No [a]	Yes DOS
User's guide available	Yes	Yes	Yes	No	Yes
Source code available	Yes	Yes	Only MINEQL portion	Yes	Yes
Programming language	FORTRAN 77	FORTRAN 77	FORTRAN 77	BASIC	FORTRAN 77 or BASIC

[a] The current release, WHAM-VI, does have a user interface.

organic matter and particulate chemistry makes it impossible to determine a single set of parameter values that will work for all conditions. The calibration of WHAM to a variety of metals and for organic matter from a variety of sources is perhaps the best that could be hoped for in establishing a universal database.

Comparison of individual models is further made difficult by the fact that most models give the user the ability to customize a given simulation, including the choice of chemical components (and therefore the reactions that will be simulated), and different formulations for a particular process (such as adsorption). The word "model" is ambiguous in this sense because each of these computer "models" (i.e., the computer code) can be used to simulate different conceptual "models" (i.e., specific theories of ion adsorption or organic matter complexation). Any of the models (computer codes) listed in Table 4-1, for example, could be used to develop appropriate descriptions of metal speciation, including complexation with natural organic matter. The fact that one of them (WHAM) has a unique description of organic matter complexation does not make it inherently superior to other models. Any of these models (computer codes) could be used to simulate the chemical and electrostatic processes that determine metal complexation with natural organic materials. The advantage to WHAM in this regard is that it has been designed to simplify and standardize the representation of natural organics and their effects on metal chemistry. For a nonexpert user, these features may be an advantage because fewer critical decisions must be made to use the model appropriately. An expert user, on the other hand, may be frustrated by the lack of generality. Ultimately, there is no simple way to determine which model is best because any of these models can be used effectively.

CHAPTER 5

Bioaccumulation and Toxicity Models

Introduction

The primary use of predicted environmental concentrations (PECs) that are evaluated with a fate and transport model is to compare them to the predicted no-effect concentration (PNEC). The PNEC may be derived from an effect concentration divided by an appropriate assessment factor, or it may be set equal to a statistically based water-quality criterion (WQC). In some instances, it may be desirable to relate the predicted water concentration and/or sediment concentration to a tissue residue level in an aquatic organism. The simplest way to do this is to multiply the predicted water or sediment concentration by a bioconcentration factor (BCF = biota concentration / water concentration in water-only exposures), bioaccumulation factor (BAF = biota concentration /water concentration, with accumulation from all routes of exposure, including water and diet), or biota–sediment accumulation factor (BSAF = biota concentration / sediment concentration), as appropriate, to predict the biota accumulation level. The magnitude of these factors is frequently used in screening-level evaluations to assess the existence of hazardous conditions or to decide on the need to conduct more rigorous evaluations (Chapman et al. 1996). Although BCFs, BAFs, and BSAFs are simple to apply, their use ignores the fact that many metals are essential for organism health (Kieffer 1991), and their level of accumulation is often regulated by the organism as a means of satisfying this requirement (Chapman et al. 1996). Thus, for metals such as copper and zinc, active uptake and elimination processes may govern the degree of accumulation by an organism. This could lead to a high BCF during periods associated with low exposure levels. The resulting elevated BCF is not indicative of significant risk to the organism and is not necessarily applicable to higher concentration ranges. More likely, it will reflect a desirable response by the organism to maintain the necessary internal levels of these essential metals.

Other more mechanistically based models are also available as an alternative to the use of BCFs and BAFs to predict the bioaccumulation and toxicity of chemicals. This section reviews the current status of several of these types of models and identifies areas where further development is needed in order to enhance their applicability to metals. The first type of model, generally referred to as a "bioaccumulation model," has historically been more commonly applied to the analysis of organic chemicals than to metals. Some of the more well-known models will be introduced in this section, and their application to metals discussed. Those readers who are interested in a further description of these types of models and the underlying principles are referred to an excellent discussion of this subject by Gobas and Morrison (2000). Models that have been proposed to predict the toxic effects of metals to aquatic life also will be reviewed in this chapter. Particular attention is given to the Biotic Ligand Model (BLM), a recent addition to the suite of models available for use in risk assessments. It is being developed specifically for the purpose of predicting the acute toxicity of dissolved metals in the water column. Finally, this section will include a discussion of acid-volatile sulfide (AVS) and simultaneously extracted metal (SEM) and how consideration of these variables can be used in the assessment of metal toxicity in sediments. The status of a model that is under development to predict AVS and SEM levels in sediments will be described.

It is noted at the outset that not all classes of accumulation and toxicity models will be reviewed in this section. Specifically, the review does not consider the more specialized fate, transport, and bioaccumulation models that are applicable to metals such as mercury and selenium. Such models are relatively complex due to the necessity of including processes such as methylation and demethylation reactions or, for metal species having a relatively high vapor pressure, volatilization (Masscheleyn and Patrick 1993; Zillioux et al. 1993).

Review of Bioaccumulation Models

The accumulation of toxic contaminants by aquatic organisms is typically viewed as a dynamic process that depends on direct uptake from the water, food ingestion, and depuration (from back diffusion, urine excretion, and egestion of fecal matter) and metabolic transformation of the contaminant within the organism. For phytoplankton and possibly other lower trophic-level animal species, direct uptake from the water is described by diffusion of the contaminant through cell membranes. For fishes and other higher trophic organisms, diffusion (e.g., through gill membranes or dermal layers) and food ingestion may both play important roles.

Several bioaccumulation models (Thomann et al. 1974; Thomann 1977, 1978; Thomann and Connolly 1984; Barber et al. 1991; Thomann et al. 1992a, 1992b; Gobas 1993; Park 1998) have been developed over the previous 30 years to describe the processes of contaminant uptake, depuration, and transformation in aquatic organisms and contaminant transfers through aquatic food webs. Overall, the models are similar in their construction and reflect a cross-fertilization of ideas among investigators (see comparison of Thomann and Gobas models in Burkhard 1998). For this reason, a general equation for bioaccumulation and the general structure of bioaccumulation models for aquatic food web models will be presented first. The application of bioaccumulation models for hydrophobic organic chemicals (HOCs) and metals will then be discussed. Recommendations for future development are presented in Chapter 6.

General equation for bioaccumulation

Model equations for the uptake and release of contaminants are often written in terms of µg contaminant per g organisms (v), where organism weight is expressed in terms of wet weight or lipid content (Thomann et al. 1992a). The general form of bioaccumulation equations is given below:

$$\frac{dv_i}{dt} = k_{ui}C_d - k_{bi}v_i + \sum \alpha_{ij} I_{ij} v_j - [k_e + k_m + k_g] v_i \quad (5\text{-}1),$$

where v_i is the concentration of the chemical in organism i (µg contaminant/g organism i), t is time, k_{ui} is the diffusive uptake rate of dissolved contaminant from the water and into the organism (L/g organism i/day), C_d is the "truly" dissolved contaminant concentration (µg contaminant/L) and typically does not include complexed forms of the contaminant, k_{bi} is the back diffusive transfer rate of contaminant from the organism to the water (1/day), α_{ij} is the efficiency of organism i to assimilate contaminant from feeding on organism j (unitless), I_{ij} is the consumption rate of organism i on organism j (g prey/g predator/day), k_e is the excretion or egestion rate coefficient for contaminant removal from organism i (1/day), k_m is the metabolic transformation rate coefficient for contaminant in organism i (1/day), and k_g is the growth rate coefficient (1/day) and is included to account for the reduction in v_i due to the increase in the weight of the organism. As noted previously, active uptake and elimination of essential metals by organisms are processes that may also affect v_i. Although not included in the form of Equation 5-1 shown here, additional terms could readily be included to represent these processes as well. Further work is needed in this area with respect to how such terms should best be formulated for metals.

A difficulty that must be addressed in bioaccumulation modeling is that the clearance rates described above (specifically k_{bi}, k_e, and k_m) tend to be both compound and organism specific. They may sometimes be evaluated from an analysis of laboratory or field data, if the appropriate data exist. Alternatively, methods have been proposed to estimate the values of these parameters from common characteristics of the substance and organism of interest, for both organic compounds and metals (e.g., Hendriks 1995).

If contaminant transfer from the water phase is the dominant uptake mechanism (which is an appropriate assumption for phytoplankton and macrophytes), the steady-state solution of Equation 5-1 is given in terms of a BCF:

$$\text{BCF}_i = \frac{v_i}{C_d} = \frac{k_{ui}}{k_{bi} + k_e + k_m + k_g} \quad (5\text{-}2),$$

where BCF_i is the ratio of v_i/C_d for uptake of contaminant from the water phase. If removal of the contaminant by excretion or egestion and metabolic transformation are negligible and the growth of the organism is small compared to back diffusion of contaminant from the organism and into the water (k_{bi}), then BCF_i is equal to

$$BCF_i = \frac{k_{ui}}{k_{bi}} \quad (5\text{-}3),$$

where the ratio of k_{ui}/k_{bi} is related to the affinity of the chemical to partition into the organism. For HOCs, bioconcentration may be related to chemical fugacity or octanol–water partitioning. For metals, bioconcentration may be related to binding of metal to specific chemical functional groups in the organism (e.g., the metal-binding protein, metallothionein).

For strongly bound chemicals, the back diffusion of contaminant from the organism into the water (k_{bi}) will tend to be small and growth of the organism (k_g) will likely serve as the primary mechanism for reducing chemical concentrations in the organism (Thomann et al. 1992b).

For higher trophic organisms, food ingestion is also expected to be an important uptake route. At steady state, the solution to Equation 5-1 is given in terms of a BAF:

$$BAF_i = \frac{v_i}{C_d} = BCF_i + \frac{\sum \alpha_{ij} I_{ij} BAF_j}{k_{bi} + k_{ei} + k_{mi} + k_{gi}} \quad (5\text{-}4),$$

where the BAF_i is the ratio of v_i/C_d for uptake of contaminant from both the water phase and food ingestion and is dependent on BAFs of lower trophic levels.

In similar fashion, steady-state bioaccumulation of contaminant in organisms may also be expressed in terms of the BSAF:

$$BSAF_i = \frac{v_i}{r_s} = \frac{BCF_i}{K_{sw}} + \frac{\sum \alpha_{ij} I_{ij} BSAF_j}{k_{bi} + k_{ei} + k_{mi} + k_{gi}} \quad (5\text{-}5),$$

where the $BSAF_i$ is again a measure of uptake of contaminant from both the water phase and food ingestion but is expressed in terms of the contaminant concentration in the sediment (r_s) in µg/g sediment, and K_{sw} is the sediment–water partition coefficient (typically in units of mL/g).

For HOCs with octanol–water coefficients greater than about 10^5 (log K_{ow} > 5), field observations for fish indicate that BAF values are about 4 times greater than BCF values (Connolly and Thomann 1992). This indicates that higher trophic-level organisms are not in equilibrium with the dissolved contaminant concentrations. Because no evidence exists for active transport of HOCs into organisms, Gobas et al. (1993) and others have hypothesized that the digestion and absorption of food in the gastrointestinal tract (GIT) of higher organisms causes the fugacity (or activity) of the contaminant in the unabsorbed food to increase. Passive diffusion of contaminant from the unabsorbed food and through the GIT membrane then is believed to result in a higher accumulation of the contaminant in higher trophic-level organisms. This results in biomagnification of HOCs as contaminated food is passed through the food chain.

For metals, food ingestion can also be a significant pathway for accumulation in aquatic organisms (Thomann et al. 1974, 1995; Fisher and Wang 1998). This is particularly true for metals such as zinc, cadmium, copper, and mercury, which induce the production of the metal-binding protein, metallothionein (Thomann et al. 1995). Once accumulated by organisms, metals are typically bound strongly to protein or sulfur groups. It has been reported that these silver complexes may be less available to predators and thus less likely to be transferred to higher trophic levels than are other forms of ingested metal (Fisher and Wang 1998). This finding requires confirmation and further testing for metals in general. Metals may also be internally sequestered by the formulation of calcium phosphate granules, and this also reduces their transfer to higher trophic levels (Nott and Nicolaideau 1990).

To date, none of these metal-specific processes have been explicitly included in any of the currently available bioaccumulation models that are reviewed herein. However, this is an active area of research (e.g., Wallace et al. 1998, 2003; Wallace and Luoma 2003). As ongoing efforts continue to elucidate the underlying processes in further detail, it is expected that they will begin to be represented in the available models.

Bioaccumulation models

The bioaccumulation frameworks incorporated in many of the currently available bioaccumulation models are similar in form and in some cases have evolved from the early work of Thomann and co-workers (1974, 1977, 1978). Thomann proposed a generalized, unified framework for modeling the bioaccumulation of chemicals in complex food webs. This model was subsequently applied to the analysis of polychlorinated biphenyls (PCBs) in the ecosystem of the Hudson River estuary (Hydroscience 1978, 1979). This type of multicompartment food chain model has continued to be refined and has since been routinely applied to evaluate the bioaccumulation of contaminants in fish that feed on lower-level trophic organisms (Thomann et al. 1974; Thomann and Connolly 1984; Connolly 1991; Thomann et al. 1991, 1992a, 1992b; Gobas 1993). For example, a generic food-chain model proposed by Thomann et al. (1992a, 1992b) is presented in Figure 5-1. Five interactive biological compartments are considered, together with the particulate and dissolved contaminant concentrations in the water column and in sediments. In these types of models, the contaminant concentration in phytoplankton is often considered to be in equilibrium with dissolved contaminant concentrations (as described by the equilibrium relationship given in Equation 5-2). The accumulation of contaminant in higher trophic organisms is dependent on both diffusive transfer (e.g., through gills) and feeding as described in Equation 5-1. Here, zooplankton obtain their food from the ingestion of phytoplankton, benthic invertebrates obtain contaminant through the ingestion of contaminated sediment particles and/or from phytoplankton and detrital matter at the sediment-water interface, forage fish feed on zooplankton and benthic invertebrates, and piscivorous fish feed primarily on forage fish.

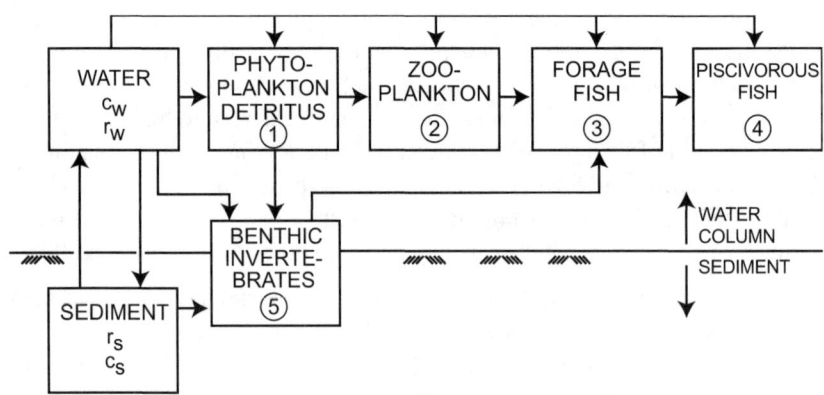

Figure 5-1 Schematic of a 5-compartment generic food web model (Adapted from Thomann et al. 1992a. Copyright 1992 CRC Press; reprinted with permission.)

For specific model applications, feeding patterns, ingestion rates (I_{ij}), growth rates (k_g), and egestion rates (k_e) are determined from bioenergetic models of energy flows through food chains and/or from stomach content, fish growth, and fecal matter production data. Because age may play an important role in describing feeding patterns and in determining the accumulation of contaminant, a further breakdown in age classes may be required (e.g., see the schematic of model compartments for age-dependent accumulation of PCBs in striped bass for the Hudson River [Thomann et al. 1991] presented in Figure 5-2). Other model parameters for diffusive uptake and backward diffusive transfer (k_{ui} and k_{bi}), assimilation efficiencies (α_{ij}), and the metabolic rate coefficients (k_m) are usually determined by model calibration to laboratory or field data.

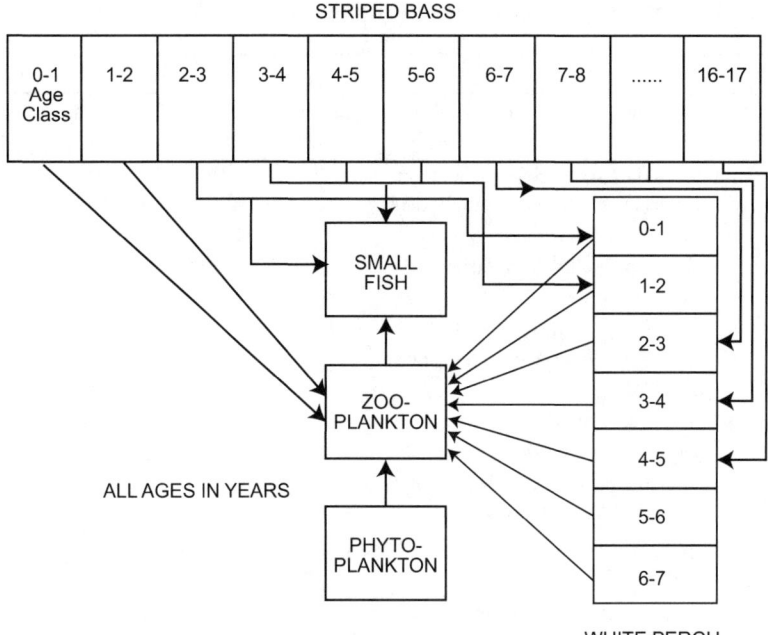

Figure 5-2 Age-dependent striped bass model for the Hudson River estuary (Adapted from Model of the fate and accumulation of PCB homologues in Hudson Estuary. Thomann RV, Mueller JA, Winfield RP, Huang CR. 1991. *ASCE J Environ Engr*. Reproduced by permission of the American Society of Civil Engineers.)

Application of bioaccumulation models

Bioaccumulation models that were considered in this review include the Thomann model (Thomann 1977, 1978; Thomann et al. 1984, 1992a, 1992b, 1995, 1997; Connolly and Thomann 1985), the Gobas model (Gobas 1993; Gobas et al. 1995), Aquatic Toxicity model (AQUATOX) (Park 1998), and Food and Gill Exchange of Toxic Substances (FGETS) (Barber et al. 1991; Suarez and Barber 1994). A partial listing of model applications for the four bioaccumulation models is given in Table 5-1 and shows that the models have largely been developed and applied for bioaccumulation of HOCs. Of the models, the Thomann and Gobas models have gained wider acceptance and use and gave similar results when applied in a steady-state analysis of PCBs in a Great Lakes food chain (Burkhard 1998). Burkhard's referral to these models as the Thomann and Gobas models is the nomenclature that is adopted for use herein.

In general, results from the Thomann and Gobas models confirm that concentrations of HOCs that are characterized by high octanol–water partition coefficients (i.e., log K_{ow} >5) are biomagnified at higher trophic levels (Figure 5-3). However, for less hydrophobic contaminants (i.e., log K_{ow} <5), model results and field data show that the uptake of contaminants is primarily through water (Figure 5-4a) and the

Table 5-1 Partial listing of hydrophobic organic chemical and metal applications for 5 bioaccumulation models

Model	HOC applications	Metal applications
Thomann	PCBs in the ecosystem of the Hudson Estuary (Hydroscience 1978, 1979)	Cadmium in a Lake Erie food chain (phytoplankton, zooplankton, fish, and birds) (Thomann et al. 1974)
	PCBs in the Lake Michigan trout food chain (Thomann and Connolly 1984)	Zinc, cadmium, copper, mercury, nickel, lead, chromium in two marine bivalves (Thomann et al. 1995)
	PCBs in lobster and winter flounder food chains in New Bedford Harbor (Connolly 1991)	Cadmium in rainbow trout (Thomann et al. 1997)
	PCB homologues in Hudson River striped bass food chain (Thomann et al. 1991)	
Gobas	PCBs in Lake Ontario food web (Gobas 1993; Gobas et al. 1995)	
AQUATOX	PCBs in forage fish and bass in E. Poplar Creek, TN (Park 1998)	
	PCBs in Lake Ontario food web (USEPA 2000)	
FGETS	PCBs in Lake Ontario salmonids (Barber et al. 1991)	
BASS		Mercury in everglades food web

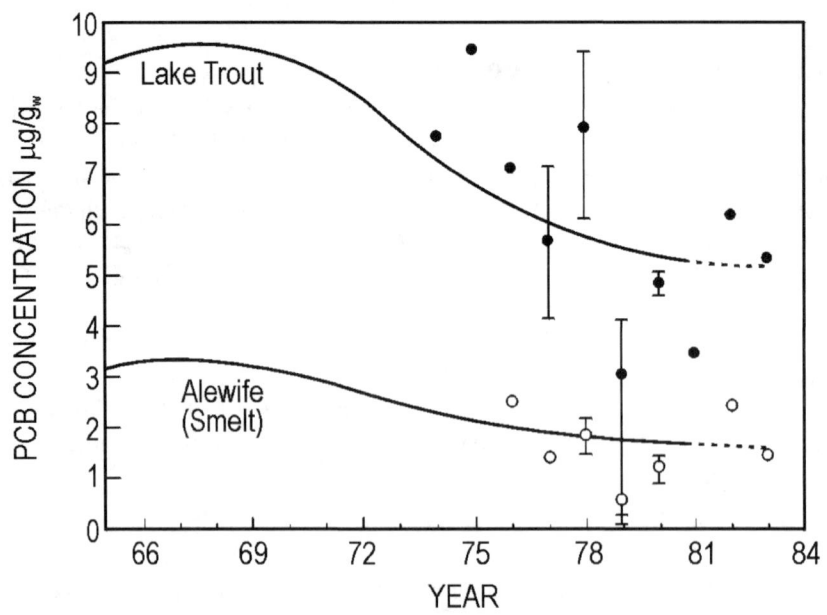

Figure 5-3 Calculated and observed time of history of total PCB in the lake trout food chain for Lake Ontario (redrawn from Thomann et al. 1992a. Copyright 1992 CRC Press; reprinted with permission.)

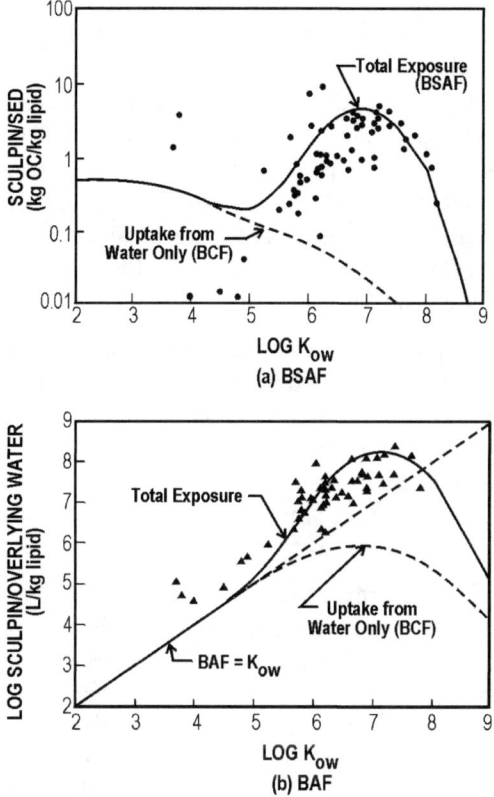

Figure 5-4 Comparison of observed BSAF and BAF data of Oliver and Niimi (1988) to calibrated sculpin model (Reprinted with permission from Thomann et al. 1992b. Copyright SETAC, Pensacola, Florida, USA)

BAF is approximately equal to the value of the K_{ow} and BCF (Figure 5-4b). For log K_{ow} greater than 7, accumulation of organic contaminants in organisms is somewhat lower. This reduction is attributed to a decrease in the assimilation efficiency (α_i) of very hydrophobic chemicals. This apparent decrease in α_i may be due to the larger chemical structure of the hydrophobic chemicals and their inability to readily diffuse through biological membranes, or it may be an artifact of specifying "truly" dissolved chemical concentrations without properly assessing chemical binding to dissolved organic carbon (DOC).

Barber and coworkers (Barber et al. 1991) have applied a bioaccumulation model of organic chemicals in fish to PCBs in Lake Ontario salmonids (alewife [*Alosa pseudoharengus*], coho salmon [*Oncorhynchus kisutch*], rainbow trout [*Oncorhynchus mykiss*], brown trout [*Salmo trutta*], and lake trout [*Salvelinus namaycush*]). The model includes passive uptake from water and from contaminated food. It considers both the biological characteristics of the fish and the physicochemical properties of the PCBs to evaluate the uptake of chemical across the gill membrane and from the gastrointestinal tract. Parameters include the gill morphometry, fish feeding, and growth rates and the distribution of PCBs across the water, lipid, and nonlipid organic material. The model was applied to laboratory whole-body accumulation data and reproduced the measured PCB accumulation quite well as a function of time, over 80 days, and versus weight. Their analysis indicated that water uptake accounted for 38% to 5%, depending on homolog group (tetra- through heptachlorobiphenyl). Thus, although uptake from food predominates, uptake via the gill was greater than previously believed, an interesting and notable finding that resulted from their analysis.

Another, somewhat simpler bioaccumulation model framework was applied to HOCs in Lake Ontario by Endicott and Cook (1994). They modeled the accumulation of 2,3,7,8-substituted polychlorinated dibenzo-p-dioxins (PCDDs) and polychlorinated dibenzofurans (PCDFs) by lake trout. The biota sediment ratios were several orders of magnitude lower than for other HOCs with similar K_{ow}s. The discrepancy could not be accounted for by chemical characteristics alone, and chemical metabolism was proposed as a possible explanation.

Examples of food web bioaccumulation model applications to metals in aquatic settings are very limited in number in comparison to applications to HOCs. In the case of metals, an important difference is that metal accumulation is strongly related to the presence of metallothioneins, a group of low molecular weight (on the order of 10,000), cysteine-rich, metal-binding proteins that have been identified in the more than 80 species of fish and invertebrates (Roesijadi 1992). For convenience, the metallothionein that is present within an organism may be considered to be comprised of two pools: a basally synthesized protein pool that is involved in essential metal regulation and an induced protein pool that is involved in metal detoxification. Metallothioneins (and associated bound metal) tend to be concentrated in the liver (or equivalent organs in invertebrates), kidneys, gills, and intestines. As discussed later in this chapter, while metallothioneins and other binding phases (e.g., calcium phosphate granules that are also believed to sequester and detoxify metals) have generally been recognized as important factors in the metabolism of metals by aquatic organisms, they have not historically been explicitly represented in applications of bioaccumulation models to metals. This is in part a result of the limited understanding of the underlying physiological mechanisms, as well as that it would add another level of complexity to the modeling endeavor. Hence, most of the applications of bioaccumulation models to metals have employed existing frameworks that were previously developed for use in modeling HOCs. Examples of selected applications to metal accumulation in marine bivalves and in rainbow trout are discussed below.

Many of the modeling investigations that have been directed at metals have focused on the bioaccumulation dynamics of individual aquatic species and simple food chains, rather than complex aquatic food webs (e.g., Reinfelder et al. 1997, 1998). This has to some degree reflected the need to gain an improved understanding of the causes of the significant differences that have been observed in metal bioaccumulation at different trophic levels, and by individual aquatic species of the same trophic level, differences that are not yet well understood. Much of this work has made use of a relatively simple but very useful kinetic model that was originally described by Thomann (1981). The kinetic model considers uptake from both waterborne and dietary routes of exposure, with the uptake rate expressed in relation to the physiological energetics of the organism of interest. In spite of it being limited in applicability to relatively simple systems, the kinetic model has proven to be of great utility in the systematic, quantitative evaluation of experimental investigations that relate to a number of important topics. These carefully designed and focused experiments have included investigations of the assimilation efficiency of metals (e.g., Wang et al. 1995, 1996; Wang and Fisher 1996a, 1996b; Reinfelder et al. 1997; Griscom et al. 2000), the trophic transfer potential of metals (Fisher and Reinfelder 1995; Fisher and Wang 1998; Reinfelder et al. 1998; Chang and Reinfelder 2000), route of exposure (Wang et al. 1997; Griscom and Fisher 2002; Griscom et al. 2002; Wang and Ke 2002) and uptake and depuration kinetics (Wang et al., 1996; Reinfelder et al., 1998; Fisher et al. 2000; Griscom et al., 2002). These datasets and related modeling analyses have added greatly to the current understanding of metal bioaccumulation. As such, they should provide an excellent foundation for ongoing metal bioaccumulation model development efforts for many years to come.

One of the earliest examples of the use of a food chain bioaccumulation model was an application to a metal, cadmium, in Western Lake Erie, USA. Thomann et al. (1974) linked this bioaccumulation model to the water column results that were provided by a cadmium fate and transport model. The food chain that was represented included phytoplankton, zooplankton, fish, and birds. While this model was generic in the sense that it did not incorporate a formulation that was specific to metals, it was a notable milestone in that it demonstrated the potential utility of this approach and it established a clear path forward in the ongoing development of food chain models.

Thomann et al. (1995) developed a bioaccumulation model for marine bivalves following the general bioaccumulation equation (see Equation 5-1). For this application, organism exposure to metals can occur through direct uptake from the overlying water or sediment pore water, and through ingestion of plankton and detritus in the overlying water or sediment (see Figure 5-5). Using this approach, Thomann et al. (1995) showed that a generalized BSAF for metals can be approximated by a simple relationship that is a function of the sediment–water partitioning, the BCF, the depuration rate, the metal assimilation efficiency from food, the bivalve feeding rate, and the organism growth rate.

The model was applied in an analysis of National Oceanic and Atmospheric Administration (NOAA) "Mussel Watch" data for two marine bivalves (*Crassostrea virginica* and *Mytilus edulis*). BSAF results (given in Figure 5-6) showed that the

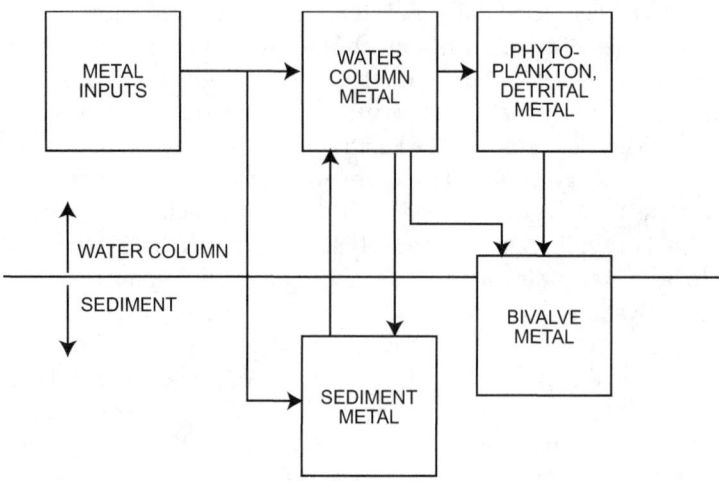

Figure 5-5 Schematic of pathways for metal uptake by bivalves (Adapted and reprinted with permission from Thomann et al. 1995. Copyright SETAC, Pensacola, Florida, USA)

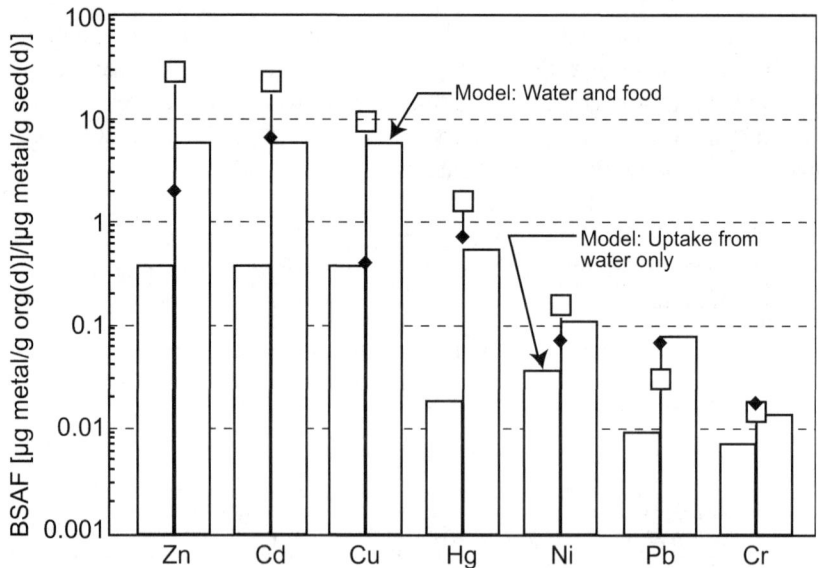

Figure 5-6 Comparison of model calibration and exposure data for *C. virginica* (squares) and *M. edulis* (diamonds). (Redrawn and reprinted with permission from Thomann et al. 1995. Copyright SETAC, Pensacola, Florida, USA)

metal accumulation from contaminated food was significant for all metals but especially for zinc, cadmium, copper, and mercury. These metals are known to induce production of metallothionein which, according to this modeling analysis, resulted in relatively high assimilation efficiency of the metal from food and relatively low depuration rates (Thomann et al. 1995).

Up to this point in the review, all the models have treated the organism as a single compartment with bulk transfer properties. Although this approach has been quite successful in describing bioaccumulation of HOCs (primarily in organism lipid compartments), it has limited application for the bioaccumulation of metals for the following reasons: First, as shown in Figure 5-7 for the accumulation of chromium in rats, metals tend to concentrate in specific organs such as the liver and kidney with time (Thomann et al. 1994). This is the result of differences in metal partitioning for the various organs (e.g., due to the concentration of metallothionein) and to specific blood perfusion rates that control the internal transfer of many chemicals (Barron et al. 1990). As a consequence, metals distribute initially into highly perfused organs (e.g., kidney, digestive tract, spleen) and then redistribute into poorly perfused organs (e.g., fat, skeletal muscle) (Barron et al. 1990). Second, dose to the target organ (and not dose to the entire organism) determines the toxicant response (Menzel 1987).

For these reasons, Thomann et al. (1997) constructed a 7-compartment PBPK model for the disposition of cadmium in rainbow trout (Figure 5-8). The model was applied to the data of Harrison and Klaverkamp (1989). Results showed that cadmium concentrations in the whole body reached steady state in approximately 50 days, but cadmium concentrations in the kidney continued to rise even during the depuration period. This was attributed to the high partitioning and retention of cadmium in the kidney by metallothionein. It is envisioned that the long range utility of the physiologically based pharmacokinetic (PBPK) modeling approach will be to evaluate the dose to the target organ, thereby improving current capabilities to predict effects. Though it differs from a PBPK model, the BLM, described in the next section, highlights the utility of predicting target organ accumulation level to predict effects.

Review of Toxicity Models for Waterborne Metals

As described previously, in the context of an aquatic risk assessment, the PECs evaluated with an environmental fate model are compared to a reference level, the PNEC. The PNEC is evaluated from an effect concentration divided by an appropriate assessment factor or, when available, a WQC for the constituent of interest. The situation is complicated by the observation that the degree of metal bioavailability and toxicity are both related to water chemistry (e.g., Pagenkopf et al. 1974; Sunda and Guillard 1976; Sunda and Hansen 1979). Formation of inorganic and organic

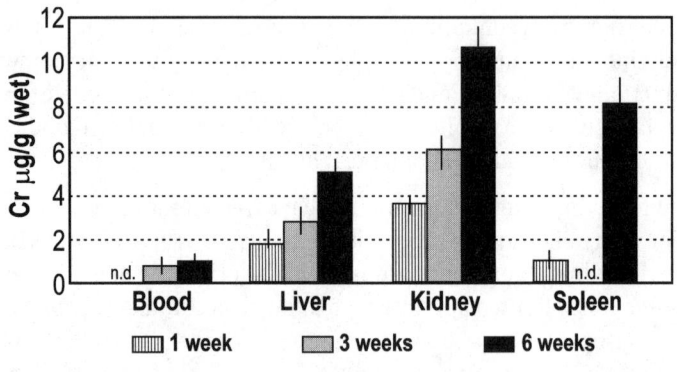

Figure 5-7 Tissue accumulation of chromium (ppm wet weight) in the rat after exposure to 100 ppm for 6 weeks. Means +/- standard error of the mean ($n = 3$ for blood; $n = 6$ for liver, kidney, and spleen) (Redrawn from *Toxicology and Applied Pharmacology*, Volume 128, Thomann et al., pages 189–198, Copyright 1994, with permission from Elsevier.)

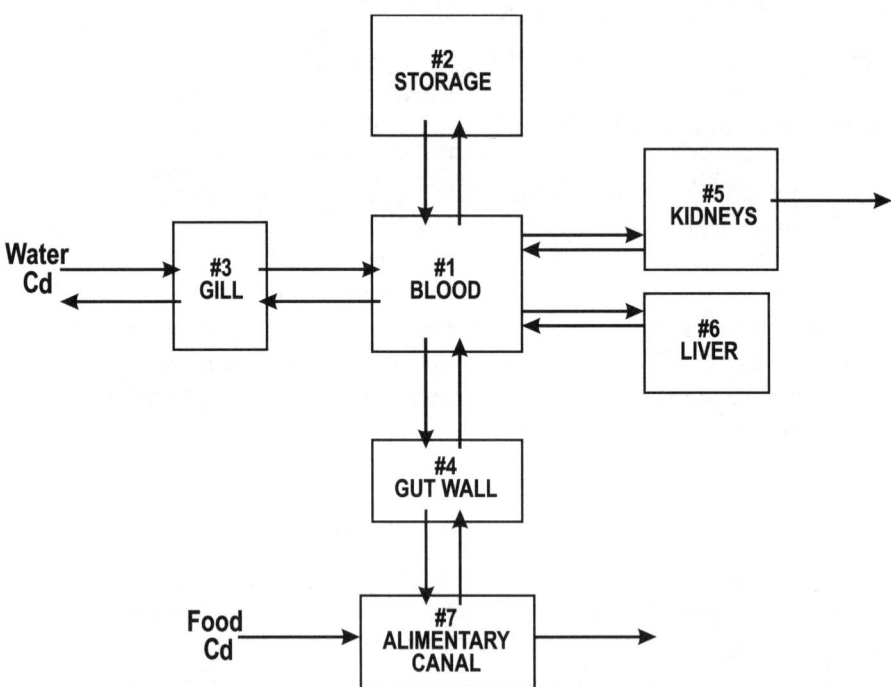

Figure 5-8 Schematic of 7-compartment pharmacokinetic model for cadmium in rainbow trout (Redrawn with permission from Thomann et al. 1997. Copyright SETAC, Pensacola, Florida, USA.)

metal complexes as well as sorption on particle surfaces have been shown to reduce toxicity. As a result, metal toxicity expressed as total or dissolved concentrations can be highly variable, depending on ambient water chemistry, while free metal ion concentrations tend to provide a much better indication of bioavailability and toxicity (Sunda and Guillard 1976). Erickson, Benoit, and Mattson (1996) conducted experiments with fathead minnows, over a wide range of water quality characteristics, to demonstrate these effects and proposed a chemical equilibrium-based model that could be used to explain the results. Allen and Hansen (1996) showed how metal speciation would be expected to affect toxicity and predicted the range of effects on copper toxicity using the site-specific water quality characteristics of a variety of water bodies.

The BLM recently was developed to incorporate metal speciation and the protective effects of competing cations into predictions of metal bioavailability and toxicity (Di Toro et al. 1997, 2001; Paquin et al. 1999; Santore et al. 2001). The BLM is based on a conceptual model similar to the gill surface interaction model proposed by Pagenkopf (1983). The BLM incorporates a version of the Chemical Equilibrium in Soils and Solutions model (CHESS) (Santore and Driscoll 1995) that has been modified to include the chemical and electrostatic interactions that are represented in the Windermere Humic Aqueous Model (WHAM) (Tipping 1994). Metal toxicity is simulated as the accumulation of metal at a biologically sensitive receptor, the "biotic ligand," which represents the site of action of acute metal toxicity (Figure 5-9). However, inorganic and organic ligands can also bind metal, thereby reducing accumulation at the biotic ligand. By incorporating the biotic ligand into a chemical equilibrium framework that includes aqueous metal complexes, the relation between free metal ion concentrations and toxicity is an inherent feature of the model. Described in this way, the BLM has obvious similarities to the free-ion activity model (FIAM), which was first formally described by Morel (1983). Campbell (1995) provides a detailed review of FIAM, and Paquin, Gorsuch, et al. (2002) describe the evolution of the BLM and FIAM and discuss their similarities and differences.

An important feature of the BLM framework is that it incorporates the competitive effects of other cations that interact with the biotic ligand to mitigate toxicity. For example, at a fixed free ion concentration, as hardness increases, the increased Ca^{2+} competes with the free metal for binding sites at the fish gill. A higher free metal concentration is therefore required to achieve the same toxic effect as Ca^{2+} increases. It is for this reason that increased hardness reduces the toxicity of some metals. It has been shown that the BLM can effectively account for reduction in metal toxicity due to elevated levels of hardness cations (Meyer et al. 1999).

The BLM has been developed using published information on metal toxicity and biotic ligand accumulation as a function of water chemistry. The most comprehensive dataset compiled to date for use with the BLM is for copper toxicity to fathead minnows (*Pimephales promelas*) (Erickson, Benoit, and Mattson 1996; Erickson

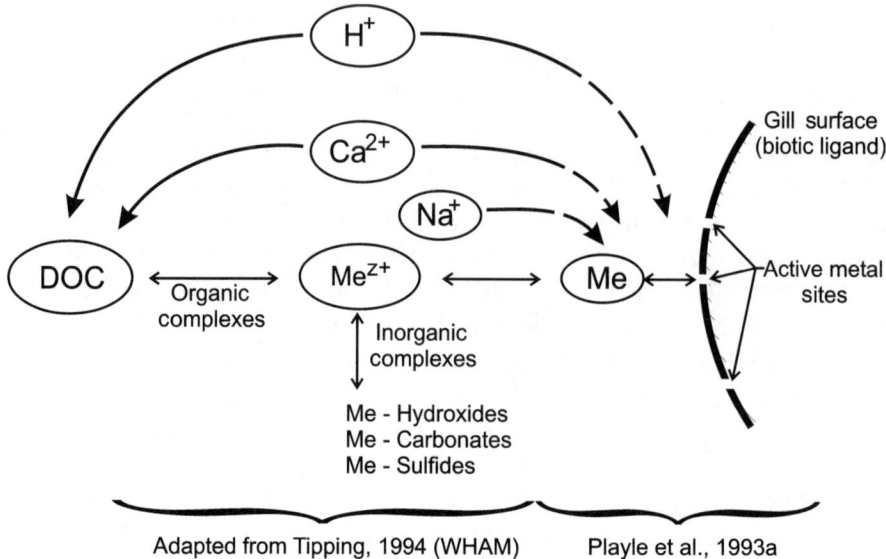

Figure 5-9 Conceptual diagram of BLM (After Pagenkopf 1983)

Benoit, Mattson et al. 1996). The biotic ligand for fathead minnows has been identified as an enzyme that is involved with the active uptake of sodium at the gill. Copper accumulation at the gill has been associated with decreased plasma sodium concentrations due to interferences with this uptake process (Playle et al. 1992). The complexation of copper at the gill in the BLM has been calibrated to measurements of copper accumulation over a wide range of water quality conditions (Playle et al. 1992, 1993b). Additionally, MacRae and coworkers (MacRae 1994; MacRae et al. 1999) established a dose–response relationship necessary to determine the biotic ligand LA50 in rainbow trout, the lethal accumulation level resulting in 50% mortality. In the BLM, metal toxicity is defined as the metal concentration in the water that is necessary to result in metal accumulation at the biotic ligand equal to the LA50. While others have developed models capable of predicting metal bioaccumulation on the gill in short-term exposures (Playle et al. 1993a, 1993b), the BLM makes use of this capability to predict toxicity.

The ability of the BLM to predict acute copper toxicity to fathead minnows was tested against measured toxicity datasets from static exposures in laboratory and natural waters over a wide range of water chemistry characteristics (Dunbar 1996; Erickson, Benoit, Mattson 1996; Erickson, Benoit, Mattson et al. 1996; Diamond et al. 1997). These datasets established that increases in pH, DOC, alkalinity, and hardness led to reduced acute copper toxicity (i.e., elevated LC50 values for copper). A comparison of predicted versus measured toxicity shows favorable results (Figure 5-10, upper panel). Nearly all predictions are within a factor of 2 of measured values over a large range in measured toxicity values (greater than 2

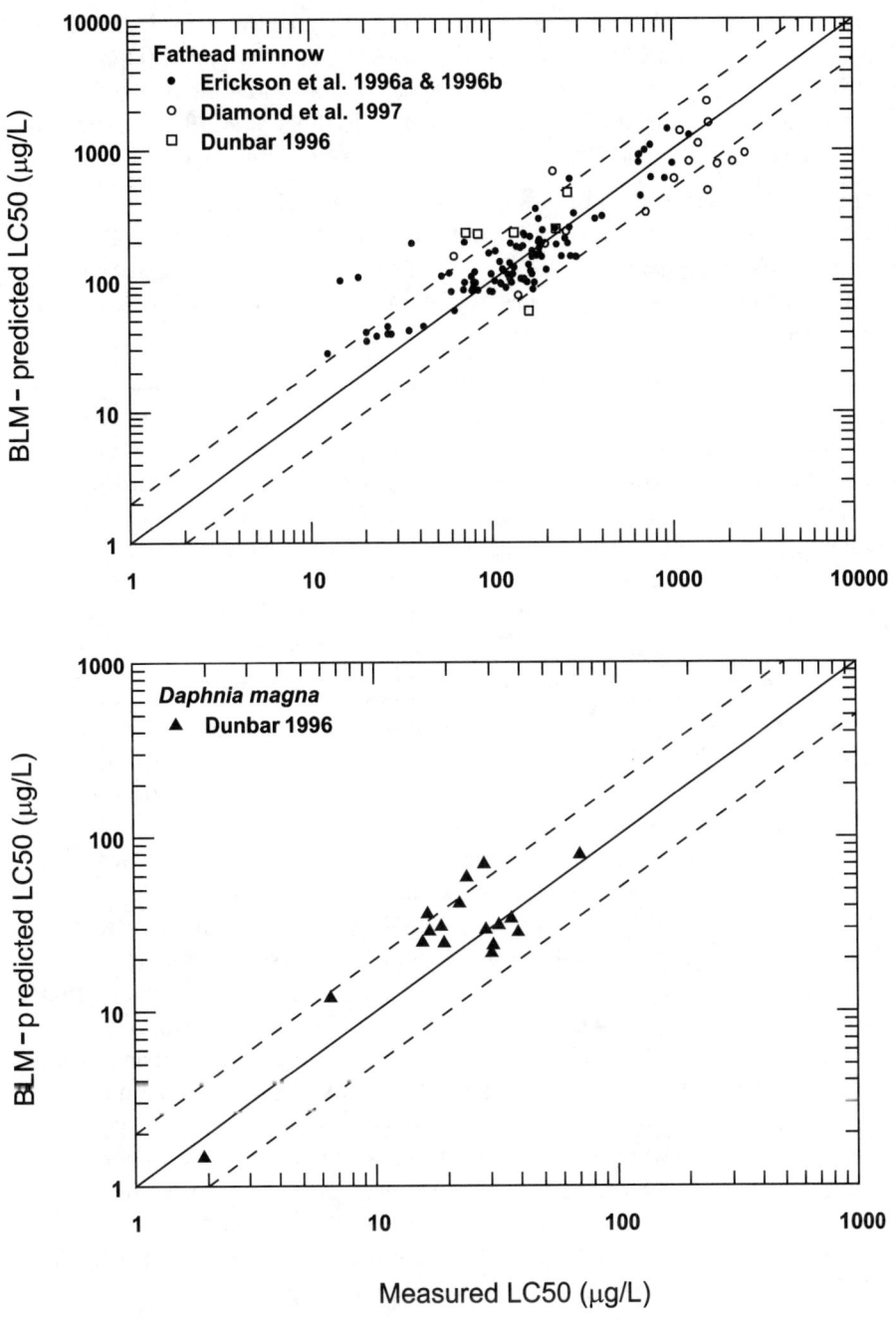

Figure 5-10 BLM-predicted LC50 versus observed LC50 for copper

orders of magnitude). Similar results have been obtained for preliminary analyses of copper toxicity data for *Daphnia pulex* (Figure 5-10, lower panel) (Dunbar 1996), and for predictions of silver toxicity data for fathead minnows and rainbow trout (Figure 5-11, upper panel) (Davies et al. 1978; Bury et al. 1999; Davies 1997; Forsythe et al. 1997) and *Daphnia magna* (lower panel) (Bills et al. 1997).

The technical basis for the BLM was presented to the Science Advisory Board (SAB) of the U.S. Environmental Protection Agency (USEPA 1999a). SAB recommendations with regard to future development needs have been released (USEPA 2000), and experimental and model development efforts are in progress to address these needs.

Other applications of the BLM framework have recently been reported. McGeer and coworkers (2000) provide a mechanistically based rationale for a variation of the BLM developed for silver, relating the biotic ligand-binding constant to the inhibition of the activity of Na^+, K^+-ATPase, an enzyme that is associated with the active uptake of sodium from water. More recently, the BLM has been applied to other organisms for copper, as well as to other metals including zinc. BLM development efforts are also in progress for cadmium, nickel, and lead as well (e.g., De Schamphelaere and Janssen 2002; De Schamphelaere et al. 2002; Heijerick et al. 2002a, 2002b; Santore et al. 2002; Paquin, Gorsuch et al. 2002). Paquin, Gorsuch et al. (2002) provide a comprehensive overview of the BLM development efforts.

A number of models have also been proposed for use in predicting the survival time of aquatic organisms exposed to a stressor. While many of these models are not available in the form of a computer program that is distributed for general use, the equations that are used are described in detail. A brief overview of the types of models that have been proposed is presented here. Mancini (1983), in one of the earliest attempts at relating the accumulation level of a metal stressor to effects, developed a model that was applied to zinc. This model represented the whole body uptake and depuration of zinc as a first-order process. It was used to simulate the time until a critical body burden of zinc was achieved. The model was applied to both steady-state and time variable conditions, with the time to mortality corresponding to the "equivalent mortality dose." Connolly (1985), using a similar approach to that of Mancini, also related whole body metal accumulation to effects. However, neither Mancini nor Connolly considered metal speciation. Further, their use of whole body accumulation rather than accumulation at the site of action of toxicity, while a useful step in the evolution of models that could be used to predict toxicity levels, made it difficult to relate the model results to a specific mechanism of toxicity. In contrast to one of the fundamental premises of the BLM, Connolly posited that it was not necessary to predict metal accumulation at the site of action of toxicity. This would be the case only if the accumulation level at the site of action of toxicity was always a fixed multiple of the whole body concentration.

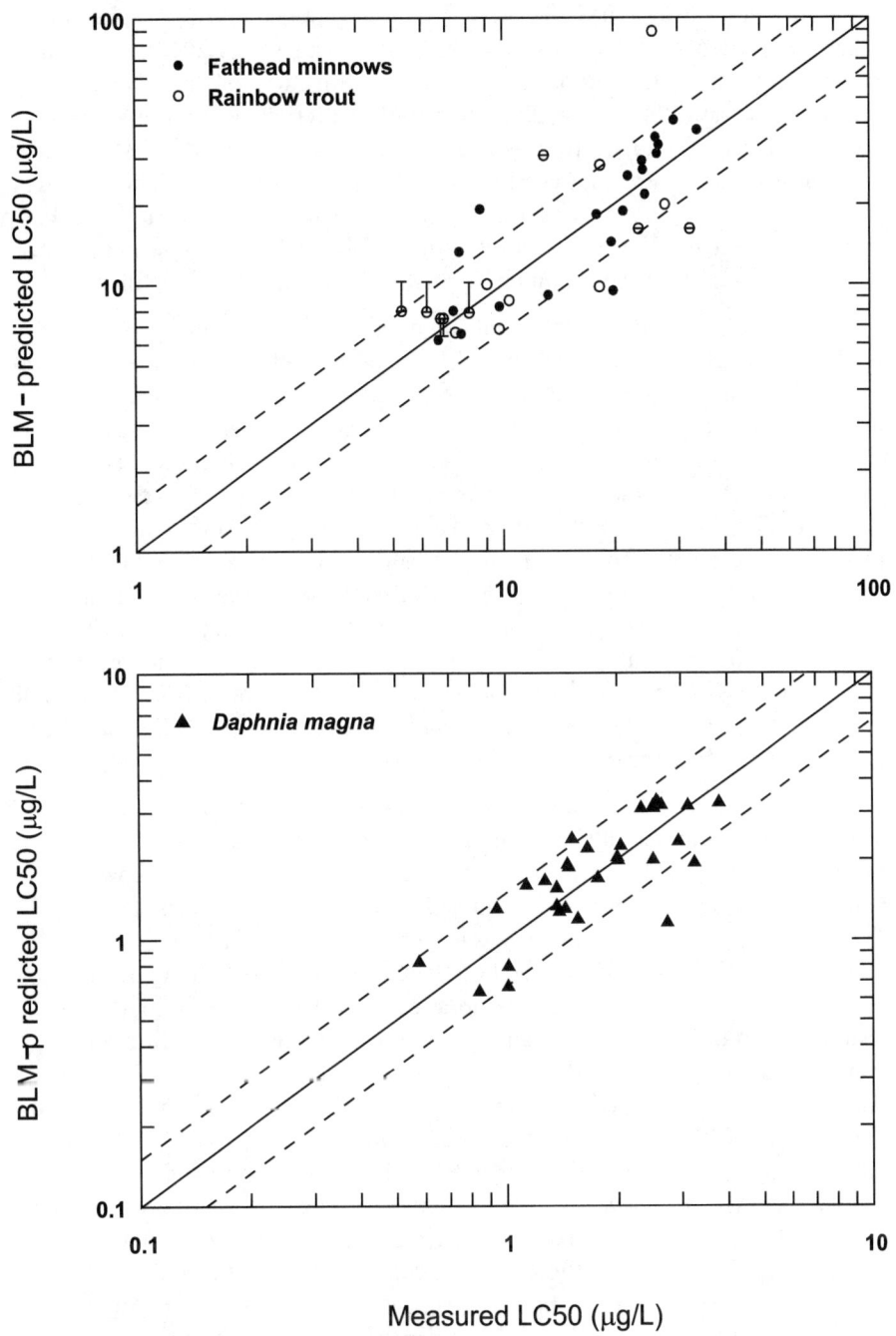

Figure 5-11 BLM-predicted LC50 versus observed LC50 for silver

The Constant Body Residue (CBR) model, typically applied to narcotic compounds, is based on the concept that the CBR associated with an effect is constant over a range of exposure times, compounds, and species (McCarty 1987; McCarty et al. 1993). Verhaar and coworkers (1999) proposed an alternative LC50 versus time model for reactive and receptor-mediated toxicants that implies that the CBR is not necessarily constant. Rather, the critical area under the curve (CAUC) of the body residue versus time curve is constant. The result is that even after steady-state body burdens are reached, the LC50 continues to decrease with an increase in exposure duration. This model is not uniquely defined for all organisms or chemicals.

Meyer et. al. (1995) compared the Mancini uptake and depuration model to a model that is based on the assumption that the product of the exposure concentration (c) and duration (t) is constant (ct = constant) and to Breck's damage-repair model (Breck 1988). It was concluded that, although the underlying mechanisms were not well represented in these models, they were still potentially useful tools for analysis purposes in some applications. Subsequently, Marr and coworkers (1998) applied a one-compartment uptake-depuration framework to model the toxicity of cobalt and copper, both independently and as a mixture, to rainbow trout. The computed incipient lethal level (ILL) was used as the indicator of organism exposure and effects (Mancini 1983 as adapted by Meyer et al. 1995 to consider accumulation at the site of action) and speciation was considered in the interpretation of the results, though not in the computation of the ILL. It was found that assuming the toxicity was additive could lead to the overprediction of toxicity in cobalt–copper mixtures.

Szumski and Barton (1983) employed a considerably different approach in proposing a mechanistically based model of acute heavy metal toxicity. This model was in some ways similar to the ion balance model described in the following paragraph, although it was based on an entirely different mechanism of toxicity. Their model, based on inhibition of respiratory function rather than ionoregulatory function, considered both metal speciation and competitive interaction of the metal at the gill membrane in fish. The interaction of the metal resulted in inhibition of carbonic anhydrase activity, thereby interfering with CO_2 excretion, and leading to death caused by the accumulation of CO_2 in the blood. Calculations were presented for copper. More recently, Roy and Campbell (1995) also considered the effects of metal speciation on the survival time of Atlantic salmon. They employed a toxic unit model to simulate the effects of aluminum and zinc over a range of pH levels. Zinc and aluminum were considered both independently and as a mixture, based on the principle that the toxicity of a mixture can be approximated by computing the sum of the concentrations of the individual metals, expressed as fractions of their respective LC50s. Proton competition was found to partially account for the residual dependency of toxicity on pH that was observed once free ion activity had been considered. This model did not consider either metal accumulation or the mechanism of toxicity.

Most recently, Paquin, Zoltay, et al. (2002) proposed an Ion Balance Model (IBM) that may be used to evaluate the survival time of aquatic organisms exposed to metals. It was applied in the analysis of data for rainbow trout exposed to silver. The model framework is similar in some ways to a physiologically based pharmacokinetic (PBPK) model, but not entirely. While the model is founded upon a physiologically based, 4-compartment representation of a fish and includes accumulation of the metal at the site of action of toxicity, it differs from a conventional PBPK model in that it does not compute the internal translocation and ultimate distribution of the stressor, in this case silver, over time. Rather, the concentration of silver at the site of action, as calculated using the previously described BLM, is used to evaluate the degree of effect of silver on the mechanisms of toxicity that are operative. The mechanisms considered include the inhibition of the active uptake of sodium and, at sufficiently high silver levels, the stimulated loss of sodium by passive diffusion. It then accounts for the subsequent impact of these changes in uptake and loss of sodium by keeping track of the cumulative damage to the fish, as manifested by loss of sodium from the internal fluid compartments. Survival time corresponds to the time when the cumulative effect is a fixed degree of loss of sodium, assumed to be 30%, from the primary vascular system.

Modeling Metal Toxicity in Sediments

The ability to assess the toxicity of nonionic organic chemicals and metals in sediments has evolved quite rapidly over recent years, with significant advances having been made since the mid-1980s. Although there are differences in the overall approach, many of the advances in how to assess sediment metal toxicity have resulted from the somewhat earlier efforts to develop sediment quality guidelines and criteria for neutral organic chemicals. As a result, this section will include a discussion of the underlying principles with regard to assessing the potential toxicity of both organic chemicals and metals in sediments. The status of ongoing model development efforts for metals will be described as it relates to these advances.

The USEPA has recognized the importance of considering bioavailability of nonionic organic chemicals in both its WQC and equilibrium partitioning-based sediment guideline (ESG) development efforts (USEPA 1989, 1993; Di Toro et al. 1991; Ankley, Berry et al. 1996). The USEPA has similarly recognized the importance of this consideration for metals WQC (Kramer et al. 1997; Renner 1997) and metals ESGs (Di Toro et al. 1992; Ankley et al. 1993, 1994; Ankley, Di Toro et al. 1996; USEPA 1994b). The development of ESGs is based on the equilibrium partitioning (EqP) approach. This approach was originally developed for neutral organic chemicals (Di Toro et al. 1991) and subsequently extended to metals, with many of the same principles applying (Ankley et al. 1994). In either case, establishing ESGs requires a determination of the extent of bioavailability of sediment-associated chemicals. The motivation for doing so was based on the frequent observation that similar concentrations of a chemical, in units of mass of chemical per mass of sediment dry weight (e.g., micrograms chemical per gram sediment), can exhibit a range of effects in sediments from different locations. (Di Toro et al. 1991 present examples of this.) If the purpose of ESGs is to establish concentrations that apply to sediments of differing types, it is essential that the reasons for this varying bioavailability be understood and explicitly accounted for in the criteria. Otherwise the criteria cannot be presumed to be applicable across sediments of differing properties, and the results of sediment toxicity tests will not be generally applicable from one sediment to another.

The observation that provided the key insight about how to quantify the bioavailability of chemicals in sediments was that the concentration–response curve for the biological effect of concern was correlated to the interstitial water (i.e., pore water) concentration (µg/L), rather than the total sediment concentration (µg chemical/g dry sediment) of the chemical (Adams et al. 1985). In addition, the porewater effects concentration, whether for organism mortality, growth, or bioaccumulation, was essentially equal to the effects concentration in water-only exposures (Di Toro et al. 1991). For nonionic organic chemicals, the concentration–response curves correlate equally well with the sediment–chemical concentration on a sediment organic carbon basis. These observations can be rationalized by assuming that the pore water and sediment carbon are in equilibrium and that the concentrations are related by a partition coefficient, K_{oc}, as shown on Figure 5-12 (right side). The term "EqP" was adopted to describe this assumption of equilibrium partitioning. The rationalization for the equality of water-only and sediment-exposure–effects concentrations on a porewater basis is that the sediment porewater equilibrium system (right) provides the same exposure as a water-only exposure (left). The reason is that the chemical activity is the same in each system at equilibrium.

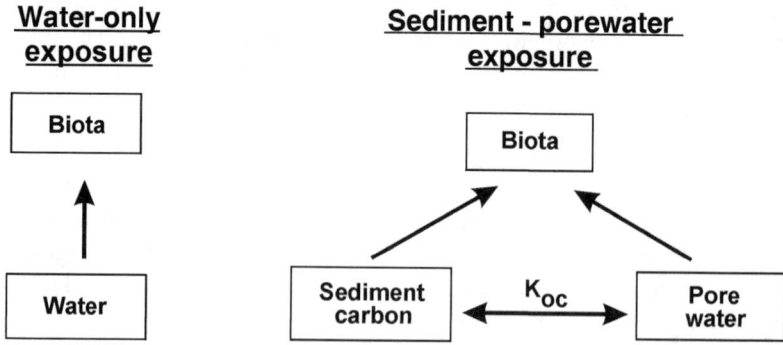

Figure 5-12 Equilibrium partitioning and water and sediment exposure. (Reprinted with permission from Di Toro et al. 1991. Copyright SETAC, Pensacola, Florida, USA)

Figure 5-13 summarizes the available data that support use of the EqP approach for nonionic chemicals. The upper panel presents the relationship of percent mortality of amphipods, exposed to 6 chemicals, to porewater concentrations that are normalized on a toxic unit basis. Typically, three different sediments were tested for each chemical. The predicted porewater toxic units represent the ratio of the measured porewater concentration to the LC50 from water-only toxicity tests. The EqP model predicts that the porewater LC50 will equal the water-only LC50 obtained from a separate water-only exposure toxicity test. A toxic unit of 1 occurs when the porewater concentration equals the water-only LC50, at which point it would be predicted that 50% mortality should be observed. The correlation of observed mortality to predicted porewater toxic units (Figure 5-13) demonstrates the efficacy of using porewater concentrations to remove sediment-to-sediment differences and the applicability of the water-only effects concentration and, by implication, the validity of the EqP model.

The lower panel of Figure 5-13 presents the percent mortality versus predicted sediment toxic units. The correlation is similar to that obtained using the porewater concentrations of the upper panel. The predicted sediment toxic units for each chemical follow a similar concentration–response curve that is independent of sediment type. These data demonstrate that 50% mortality occurs at about one sediment toxic unit, independent of chemical, species of organism, or sediment type, as expected if the EqP assumptions are correct.

It should be pointed out that the EqP assumptions are only approximately true. Thus, the predictions from the model have an inherent uncertainty. If the assumptions were exactly true, then the data on Figure 5-13 would all predict 50% mortality at one toxic unit. The results of Figure 5-13 indicate that the LC50 predictions for organic chemicals are typically good to within a factor of approximately 2. This uncertainty reflects the inherent variability of these experiments as well as the effects of other phenomena that have not been accounted for in the EqP model. This factor of 2 in uncertainty appears to be the limit of the accuracy and precision to be expected.

The EqP approach, originally developed for nonionic organic chemicals, was subsequently extended to metals, including copper, cadmium, nickel, lead, and zinc (Di Toro et al. 1992, 1999; Ankley, Di Toro et al. 1996; Berry et al. 1996). Recently, silver has been included as well (Berry et al. 1998) and testing is in progress with chromium. Briefly, the EqP methodology as applied to metals recognizes that when AVS is present in a sediment, it reacts with the SEM to form an insoluble metal sulfide that is not bioavailable. When an attempt is made to correlate organism mortality to bulk sediment metal concentration (mmol total metal/g sediment), the data do not exhibit a dose–response relationship over a range of sediment types (Figure 5-14, upper panel). When the SEM data are normalized to AVS, however, a consistent pattern emerges (Figure 5-14, lower panel). When the SEM-to-AVS ratio is less than 1, SEM < AVS, the SEM exists as the

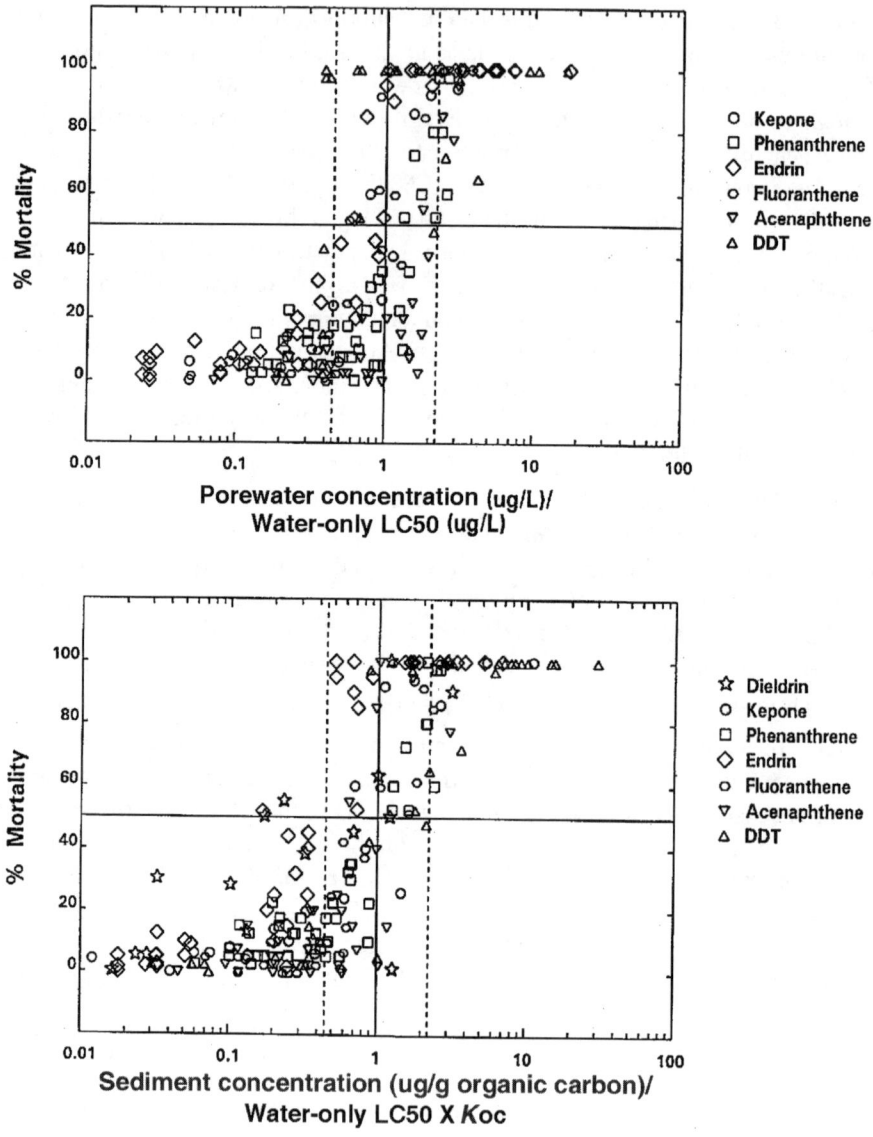

Figure 5-13 Dose–response curves based on porewater and carbon-normalized sediment concentrations

nonbioavailable metal sulfide, and toxicity is not observed. Conversely, when the ratio is greater than 1, SEM > AVS, the potential exists for dissolved metal to be present in the pore water. At this point, the presence of other binding phases, including DOC and POC, may be sufficient to preclude toxicity, and toxicity may or may not be observed.

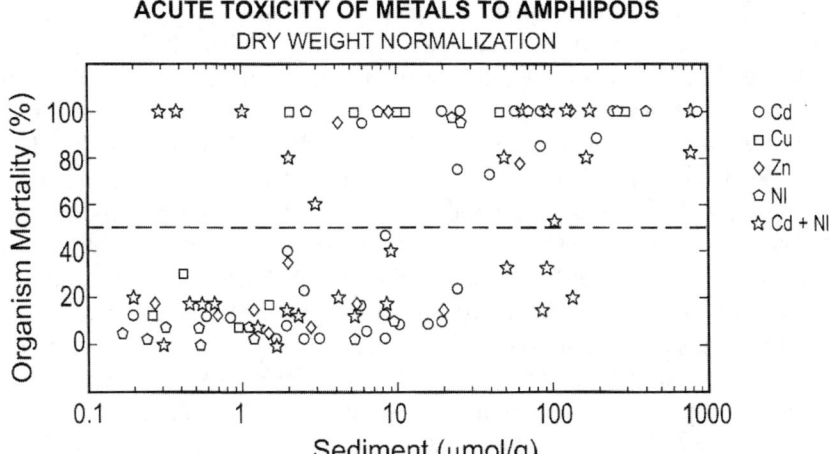

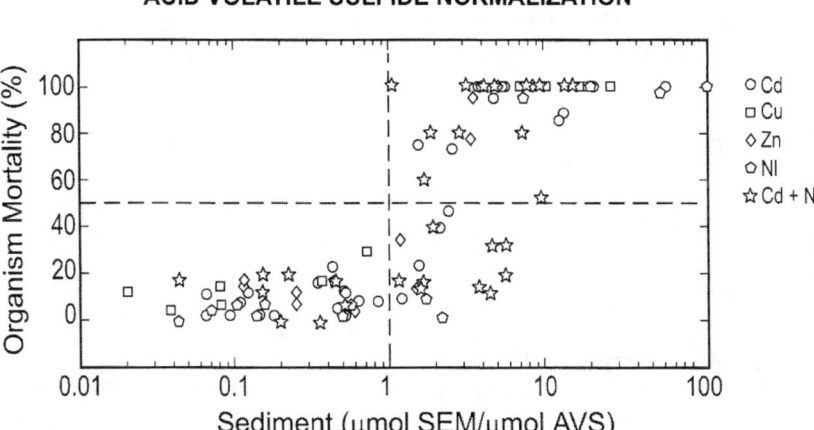

Figure 5-14 Comparison of dose–response curves for mortality as a function of bulk sediment metal concentration and SEM/AVS

Following the preceding line of reasoning, recently completed analyses suggest that mortality may be better related to the difference between SEM and AVS, normalized by the fraction organic carbon of the sediment, f_{oc} (Di Toro et al. 2003). That is, the use of the organic carbon-normalized difference between SEM and AVS [i.e., (SEM – AVS) / f_{oc}] appears to provide an improved predictor of metal toxicity in sediments. This approach promises to further reduce the uncertainty of predictions of sediment toxicity for metals, particularly for conditions where SEM > AVS.

It should be evident from the preceding discussion that the prediction of sediment toxicity will be predicated on the availability of a mechanistically based model for predicting SEM, AVS, and porewater concentrations of metals. Work of this nature is currently under way for cadmium, copper, silver, and other metals, and the initial results are promising (Di Toro, Mahony, Gonzalez 1996; Di Toro, Mahony, Hansen et al. 1996; Carbonaro 1999; Di Toro 2001). The model framework, shown in Figure 5-15, incorporates multiple sediment layers within both an aerobic surficial sediment zone and a deeper anaerobic sediment region. The model is capable of simulating the temporal and vertical variation of AVS, SEM (cadmium), and porewater cadmium in sediments. To illustrate, Figure 5-16 shows how results obtained using a high resolution version of this model (i.e., 1 mm thick sediment layers) are in good agreement with the vertical profiles of SEM/AVS that were measured in experiments with sediment cores dosed with cadmium (Di Toro, Mahony, Hansen et al. 1996). The marked gradients in the upper 20 mm of these cores reflects the combined effects of oxidation of CdS in the aerobic layer of the model and the resulting diffusive flux of dissolved metal from the sediment to the overlying water. In cases where the predicted SEM/AVS exceeds 1.0, SEM may be present in the pore water. However, as noted above, the presence of other binding phases may still be sufficient to prevent toxicity.

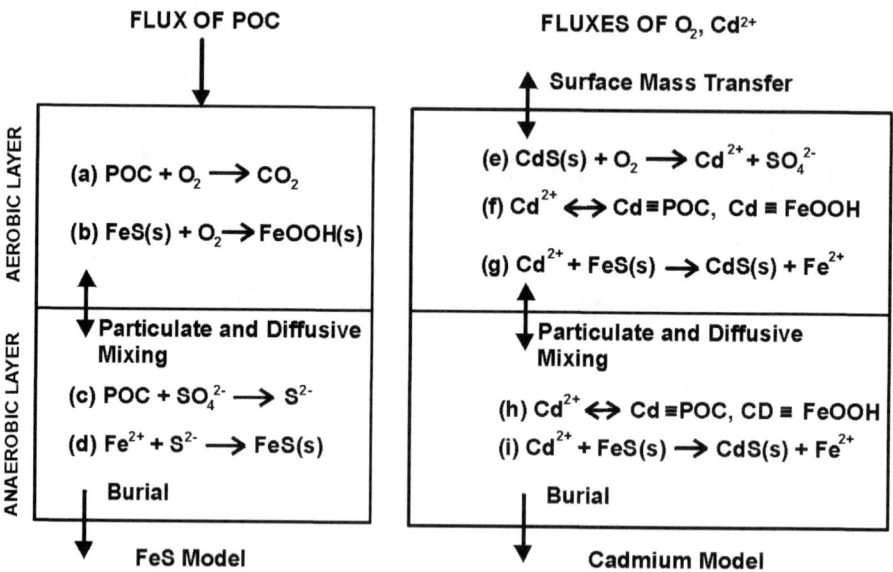

Figure 5-15 AVS and SEM model framework

Lab Colonization Experiment - SEM/AVS

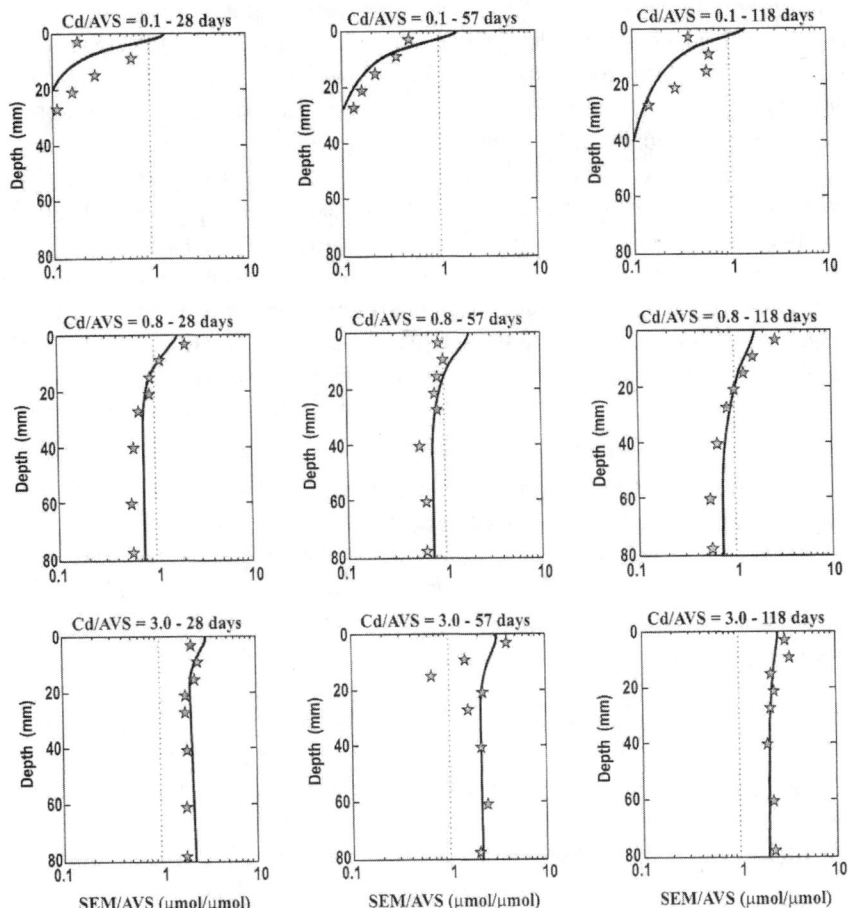

Figure 5-16 Comparison of vertical profile data to SEM/AVS predicted with model

Integration of Bioaccumulation and Toxicity Models for Metals

The previous review of bioaccumulation models discussed some of the important differences in the ways that aquatic organisms accumulate metals in comparison to organic chemicals. In view of these differences, there are some inherent limitations that should be recognized when bioaccumulation models that were developed for organic chemicals are applied to metals, with regard both to the trophic transfer potential of accumulated metals and to the ability to relate metal accumulation levels to effects. Of similar importance, the discussion of the BLM highlighted the importance of considering metal availability in the abiotic environment, and of evaluating metal accumulation at the site of action of toxicity, when assessing the potential for acute toxicity via waterborne exposure to metals. However, while much progress has been made with regard to assessing the potential for acute effects due to metals, the ability to predict the potential for chronic effects that result from longer-term exposure to metals, exposure via both waterborne and dietary sources, is less well understood, even though this consideration is important from a regulatory and scientific standpoint. Hence, there is some concern that WQC that are based on the waterborne route of exposure alone do not adequately address the potential for adverse effects that may arise via the dietary route of exposure (Schlekat and Luoma 2000). While the dietary route of exposure may result in tissue accumulation levels that are considerably in excess of homeostatically maintained basal levels of essential metals, levels that are greater than what might occur over the time frame of an acutely toxic waterborne exposure, the potential for effects related to these tissue metal levels is not considered in models such as the BLM.

When attempting to assess the potential for effects on the basis of tissue metal accumulation levels, a tiered approach might reasonably be considered. Tier I could consider measured whole body metal accumulation levels (e.g., Salazar and Salazar 1995, 1998; Salazar et al. 1996), where an attempt is made to consider the correlation of tissue metal levels to effects. The advantage of using measured whole body tissue levels is that they inherently reflect the effects of site-specific chemistry on metal bioavailability and accumulation, and they integrate the effects of varying exposure levels over time and space. A bioaccumulation model could serve as an alternative way to evaluate whole body tissue metal levels in a Tier I assessment, but such models must be used with caution to ensure that they yield reliable results. The advantage of a model would be that it can also be used to predict future tissue levels that might result from alternative remediation scenarios.

While use of total body burden (measured or modeled) offers a convenient basis for a Tier I analysis, this approach does have some limitations. One such limitation is that it may sometimes be difficult to establish cause–effect relationships when multiple stressors are present. Even when this is not the case, it also requires that an unambiguous relationship be established between a benchmark whole body tissue accumulation level and a related effect. As learned from the development of the BLM, it is useful to consider the level of metal accumulation at the actual site of action of toxicity when developing such relationships. This can be problematic if whole body tissue levels are concerned because metals that are accumulated from food and water are not typically found to be uniformly distributed throughout the organism. Rather, it is known that metals will preferentially accumulate in selected organs, often the organs involved in uptake (e.g., the gill or digestive gland) or excretion (e.g., the liver or kidney). If the accumulation level at the proximate site of action of toxicity were a fixed proportion of the whole body accumulation level, as might be the case for an organic chemical, then this consideration would be less of a concern. However, this does not appear to be the case for metals, and unfortunately, lipid normalization does not appear to resolve this problem for metals as it seems to do for organic chemicals (Phillips 1980). Alternatively, measurement of metal accumulation levels in specific organs may improve the situation, provided that a relationship between accumulation level and effect can be defined. This could be considered an appropriate approach as part of a Tier II effects assessment. PBPK-type models, which are designed to evaluate organ-specific tissue levels, might serve as an alternative way to evaluate tissue-specific accumulation levels, though their application to metals has seen limited use to date (e.g., Thomann et al. 1994, 1997; Paquin, Damiani et al. 2002).

As suggested above, evaluation of the form of the metal that is present in a tissue, either by measurement or calculation, may provide a less ambiguous indicator of the potential for effects than total metal concentration alone. The reason for not being able to make a definitive statement in this regard is that, while there are considerable data on measured total tissue levels of metals, relatively little data currently exist on the intracellular distribution of metals. Intracellular speciation of metals has been found to be important, however, with metallothionein (MT) complexes serving as one of the more important forms to be considered (e.g., Hildebrand et al. 1979; Roesijadi 1992; Mason and Jenkins 1995). Metals have also been found to concentrate within organophosphorus granules in organism tissues, and these granules are also viewed as sequestering the metal, thereby rendering them non-available for interference with other intracellular target enzymes (Coombs and George 1978; George 1982). While either of these methods of sequestration will lead to an increase in tissue metal levels, they do not necessarily result in adverse effects.

Another complication associated with using total metal levels in tissues to predict effects is that it is not necessarily the absolute level of metal accumulation that matters, but the rate of uptake of the metal (Roesijadi 1992; Hook 2001; Hook and Fisher 2002). This is believed to occur as a result of the ability of some aquatic organisms to sequester metals that enter the cell (e.g., by inducing the synthesis of MT and/or granule formation). That is, adverse effects are avoided as long as the rate of metal uptake does not exceed the rate at which the organism is able to detoxify the metal, thereby limiting the degree of increase of the concentration of bioreactive metal in the cytosol. If the rate of uptake is too high, the complexation capacity of the binding ligand (e.g., MT) may be exceeded, the bioreactive cytosolic metal levels become unacceptably high, and adverse effects may ensue. This idea, often referred to as the "spillover hypothesis" of metal toxicity, was first proposed by Winge and co-workers (1974).

While toxicity related to the increase in tissue metal concentration in an organism is mitigated through binding to certain intracellular ligands, the fact that the metal is sequestered can lead to an overall increase in tissue metal concentration over time. This increase may in turn lead to increased dietary exposure of the metal to higher-level predators. However, the specific form of the metal that is present in the prey has been shown to affect the trophic transfer potential of the metal and, presumably, the degree of effects on a higher-level predator. For example, metals that are contained in granules have been shown to resist assimilation by higher trophic-level predators (Nott and Nicolaideau 1990). At the same time, other more recent studies have shown that metals that are present as MT complexes may in some cases be available to the next higher trophic level (Wallace and Lopez 1996, 1997; Wallace et al. 1998).

Fractionation methods have been proposed for measuring the intracellular distributions of metal, and it is entirely plausible that such methods may prove to be of future use in assessing the potential for effects (e.g., Winge et al. 1974; Brown and Parsons 1978; Benson and Birge 1985; Roesijadi and Klerks 1989). As an example, one measurement scheme attempts to operationally differentiate between the particulate fraction and the low molecular weight (LMW < 3000), intermediate molecular weight (IMW on the order of 10 to 12,000) and high molecular weight (HMW > 70,000) intracellular cytosolic fractions (Roesijadi and Klerks 1989). The target enzymes that are adversely affected by accumulated metal levels are believed to be associated with the HMW fractions, and it is the "spillover" of the metal into this fraction that is expected to be most closely related to the observance of adverse effects (Winge et al. 1974; Brown and Parsons 1978). While the development of models that can perform this sort of evaluation would likely be useful, the successful application of such models in an effects assessment remains to be demonstrated. As a general rule, consideration of intracellular speciation measurements is considered to be, at least for the present, an approach that would be limited to use in a relatively advanced Tier III assessment.

CHAPTER 6

Model Selection and Future Model Development Needs

The specific objectives of this literature review were the following:

- to identify and critique candidate fate and transport models for use in evaluating exposure levels of metals in surface waters and sediments,
- to consider the utility of these models to evaluate metal bioavailability,
- to identify and critique candidate bioaccumulation and toxicity models for metals,
- to evaluate the strengths and weaknesses of these models with respect to their use with metals and metal compounds,
- to identify the most appropriate applications of existing models, given their current level of development, and
- to identify weaknesses in the available modeling frameworks and recommend ways to improve the current model capabilities.

It should be recognized at the outset that no single model formulation is universally accepted as being correct for use in all applications. Also, it has not been possible to test all of the models that have been reviewed herein to ensure that they perform correctly, so users must be vigilant in watching for problems with any model that is used. Finally, the experience of the user is likely to be at least as important a factor in the completion of a defensible model analysis, if not more so, than is the model itself. With these caveats in mind, recommendations are presented here with respect to models that may be used for fate and transport analyses and for the evaluation of bioaccumulation and toxicity. Additionally, recommendations are made with regard to how future model development efforts should be directed to best enhance the predictive capability of these models when they are applied to metals.

Model Selection

Figure 2-1 (p 7) illustrates the principal types of models that are commonly used in fate and transport analyses and aquatic risk assessments. The review did not consider in great detail the stand-alone hydrodynamic and sediment transport models that are sometimes used in advanced fate and transport analyses. However, recognizing that state-of-the-art hydrodynamic and sediment transport models are likely to be used in some situations, these models are discussed briefly. This discussion is followed by recommendations pertaining to use of fate and transport, chemical equilibrium, and bioaccumulation and toxicity models.

Hydrodynamic models

Reliable predictions of currents may be critically important to a modeling analysis because current-generated shear stresses control sediment transport processes, processes having a direct effect on the transport and ultimate fate of metals. Many fate and transport analyses incorporate relatively simple characterizations of freshwater inputs and flow patterns, and this is often sufficient for modeling purposes. Of course, verification of the flows that are assigned by comparing computed conservative tracer results to field data is a very important step in the analysis, even when this simple approach is used. There are situations, however, where a more sophisticated modeling approach is needed. For example, use of a stand-alone hydrodynamic model may be warranted in an extremely complex setting where the model results have a high degree of visibility and important long-term ecological or economic implications. The Estuary, Coastal, Ocean Model (ECOM) (Blumberg and Mellor 1987; Blumberg et al. 1993, 1999) is actually a family of state-of-the-art hydrodynamic models that would be suitable for use in this situation. ECOM employs a sophisticated turbulence closure algorithm to evaluate dispersive mixing and is capable of providing realistic time variable currents even in complex systems. Other advanced hydrodynamic models that could be considered for use include, but are not limited to, DYNHYD5 and EFDC.

Sediment transport models

As is the case for fluid transport, the fate and transport analysis will often be based on a relatively simple representation of particle dynamics in the receiving water as well. The model inputs are typically set by calibration to water column suspended solids data and long-term sedimentation rate data. There are situations, however, where a more mechanistically based sediment transport model may be warranted if reliable predictions are to be made beyond the range of model calibration conditions. For example, the Estuary, Coastal, Ocean Model–Sediment (ECOMSED) (described briefly in Chapter 3) has been developed to simulate both cohesive and noncohesive sediment transport and is capable of simulating armoring of the bed during periods of high flow. Armoring is an important process to consider, because it can have a significant effect on predicted exposure levels: When the surface of the sediment armors, the resuspension of bed sediments ceases. This in turn has significant implications with respect to predicted exposure levels. Advanced sediment transport capabilities also are available in models such as DELFT3D and EFDC.

Fate and transport models

Even when stand-alone hydrodynamic and sediment transport models are used in support of the exposure assessment, the fate and transport model is ultimately used to make the final evaluation of metal exposure levels. No single model is best for use in all cases. Although there are advantages and disadvantages associated with each of the fate and transport models reviewed herein (Table 3-1, p 16), they generally provide the analyst with useful tools for application in the exposure assessment. The following models are suggested for use in various situations:

- The simplified procedures described in Water Quality Assessment Methodology (WQAM) will serve as a useful starting point for making relatively simple calculations of exposure levels in screening-level analyses. This methodology also is suggested for use by individuals who are relatively inexperienced in working with fate and transport models.
- Simplified Lake and Stream Analysis (SLSA) is an appropriate model to use in making a more mechanistically based evaluation of metal exposure levels in a 1-dimensional (1-D) stream or river setting, or in a simple lake setting. Michigan River (MICHRIV) is essentially the same model, but has additional flexibility with respect to representing a 1-D channel that has a variable cross-section.
- Chemical Transport and Analysis Program (CTAP) is suggested for use as a steady-state model when a multidimensional representation of an irregularly shaped water body is needed. Typically, a model such as CTAP will be more time consuming to set up than a simpler model such as SLSA or MICHRIV.
- If a detailed time variable analysis of a complex setting is needed, Water Quality Analysis Simulation Program, version 5 (WASP5) or the Water Quality Analysis Simulation of Toxics (WASTOX) program should be used. These models call for a considerably higher level of experience by the user, especially if used in conjunction with a state-of-the-art hydrodynamic or sediment transport model. As an alternative, DELFT3D also provides a refined modeling system that is applicable to complex settings.
- Although not reviewed in detail here, an updated version of EFDC is being developed for use in advanced fate and transport applications. The updated version is currently scheduled for release by late 2003 or early 2004.

An important limitation of most of the fate and transport models reviewed herein is that they do not provide a very refined approach to evaluating metal chemistry. Thus, it is recommended that they be used in conjunction with a chemical equilibrium model, discussed next, for purposes of evaluating speciation and complexation reactions. DELFT3D does include a chemistry submodel that is applicable to certain "standard applications."

Chemical equilibrium models

There are many chemical equilibrium models documented in the literature (Nordstrom et al. 1979; Bassett and Melchior 1990). Of the ones reviewed herein (Table ES-2, p *xxi*), several have gained relatively widespread acceptance (most notably MINTEQA2 and MINEQL/MINEQL+) and are frequently used. Recently, the Windermere Humic Aqueous Model (WHAM) has further advanced the capabilities of chemical equilibrium models in applications to natural water systems as a result of its standardized calibration of metal–organic matter interactions to a variety of published data. The Chemical Equilibrium in Soils and Solutions (CHESS) model is a relatively new chemical equilibrium model that is also described in this review (Santore and Driscoll 1995). The Biotic Ligand Model (BLM), used to predict metal toxicity and described subsequently, is based on CHESS, modified to incorporate the metal–organic matter interaction approach employed in WHAM.

Based on a feature-by-feature comparison of the these models, many similarities and some important differences are evident. Although any of these models could be used effectively in an environmental exposure and risk assessment for metals, the determination of the "best" model to use must be made on a case-by-case basis. The choice of model will depend on what features are most important for a given application and for a given user. Consideration of the following criteria will aid in the selection of an appropriate model.

A thermodynamic database is included with the following:
- MINTEQA2, MINEQL+, and WHAM and are actively supported and updated.
- MINEQL+ is perhaps the most extensive because it includes the MINTEQA2 database plus extensions and updates.
- WHAM database is specific to metal–dissolved organic matter (DOM) interactions.

A user interface is included with the following:
- MINEQL+ has a polished interface, including versions for DOS and Windows.
- MINTEQA2 interface (PRODEFA2) is not as polished but still very powerful.
- A user interface has recently been released for the BLM, which makes use of CHESS for chemical equilibrium calculations.

With respect to inorganic speciation, differences between models may generally be resolved by user-specified modifications to the available thermodynamic databases.

A wide variety of approaches are used to represent metal–organic matter interactions, from simple interactions based on known organic acids to complex heteroge-

neous mixtures of organic functional groups with chemical and electrostatic interactions. Examples of these include the following:
- WHAM formulation is internally specified within the code.
- MINTEQA2, MINEQL+, and CHESS use a more flexible approach, and could be used to represent WHAM formulations with some effort on the part of user.
- WHAM has the best set of general purpose calibration values.

With respect to adsorption on surfaces, a wide variety of formulations are employed, from simple (partitioning) to complex surface complexation models with electrostatic and diffuse layer adsorption. The following are examples:
- WHAM is hard-wired and includes adsorption on clays and organics.
- WHAM has a very specific structure. Revisions to the code would be necessary for most modifications to this structure (other than adding simple speciation reactions to the database). Also, WHAM does not consider the solubility limitations of most metals.
- MINTEQA2, MINEQL+, and CHESS have a generic structure and are easily modified for simulating different systems without altering the code. They could be used to duplicate WHAM functionality for metal–DOM interactions, but this would require some effort on the part of the user.
- MINTEQA2, MINEQL+, and CHESS have been used to incorporate chemical equilibrium reactions within other modeling frameworks (such as fate and transport, nutrient cycling, and BLMs).

Finally, it is noted that the Nonideal Competitive Adsorption model (NICA) (Koopal et al. 1994; Benedetti et al. 1995; Kinniburgh et al. 1996; Temminghoff et al. 1997; de Rooij et al. 1999), described briefly in Chapter 4, is another recently developed chemical equilibrium model that could be considered for use.

Bioaccumulation and toxicity models

The bioaccumulation models reviewed herein were summarized previously in Table 5-1 (p 68). These models have been developed over the previous 30 years to describe the processes of contaminant uptake, depuration, and transformation in aquatic organisms and contaminant transfers through aquatic food webs. Of these models, the Thomann and Gobas bioaccumulation models have been widely applied and are considered to be most generally accepted by the scientific community. These two models yielded similar results when compared in a steady-state analysis of polychlorinated biphenyls (PCBs) in a Great Lakes food chain (Burkhard 1998). Compared to hydrophobic organic chemicals (HOCs), examples of bioaccumulation model applications to metals are much more limited in extent, and these applications have focused on bioaccumulation dynamics of individual aquatic species (e.g., Reinfelder et al. 1997). Significant differences in bioaccumulation at different trophic levels and by individual species of the same trophic level have been observed and are not well understood at this time. As a result, further development is needed in this area. Even so, the current version of either the Thomann or Gobas models will provide the analyst with useful tools for assessing metal bioaccumulation. The Thomann model has been more widely applied to metals to date.

Metal bioavailability and toxicity have long been recognized as being a function of water chemistry (e.g., Pagenkopf et al. 1974; Sunda and Guillard 1976; Sunda and Hansen 1979). The formation of inorganic and organic metal complexes and sorption of metals to particle surfaces have been shown to reduce toxicity. As a result, the relationship of metal toxicity to total or dissolved concentrations can be highly variable, depending on ambient water chemistry. In comparison to these measurements, the free metal ion concentration provides a much better indication of bioavailability and toxicity (Sunda and Guillard 1976). Allen and Hansen (1996) showed how metal speciation is expected to affect toxicity and, using chemical equilibrium calculations, predicted the range of effects on copper toxicity resulting from the variation of site-specific water quality characteristics for a number of water bodies.

The BLM was developed to incorporate metal speciation and the protective effects of competing cations into predictions of metal bioavailability and toxicity (Di Toro et al. 2001; Paquin, Gorsuch et al. 2002). The BLM is based on a conceptual model similar to the gill surface interaction model proposed by Pagenkopf (1983). Although models capable of predicting metal bioaccumulation on the gill in short-term exposures have been previously developed (Playle et al. 1993a, 1993b), the BLM employs this approach in conjunction with a scheme to predict metal toxicity. While still in the process of being extended in applicability to additional organisms and metals, the BLM holds promise as a way to predict the effects of site-specific water quality on toxicity. As an example, it can be used to predict the acute effect levels of both copper (fathead minnows and *Daphnia pulex*) and silver (fathead minnows, rainbow trout and *Daphnia magna*) to within a factor of 2 over a wide range of water quality conditions. The model also has been applied in the analysis of copper toxicity data for the blue mussel, *Mytilus edulis*, a saltwater species. The USEPA is currently in the process of incorporating the BLM into an updated freshwater water quality criterion (WQC) for copper. A user-friendly program interface is also being developed as part of this effort.

A number of other toxicity models that have been used to predict the accumulation and effects of chemicals, including survival time, were also described in Chapter 5. Typically, survival time and effects are associated with the time required to achieve a critical body residue (CBR) level, or some variation thereof, in these models. While having been applied to metals in some cases, these models are best suited for applications to organic chemicals where the mode of action is narcosis. The applicability of such models for use in predicting effects due to metals is an area of active research. A somewhat different approach is used in a recently developed Ion

Balance Model (IBM), developed specifically for metals, which is also described in Chapter 5 (Paquin, Zoltay et al. 2002). The IBM uses the BLM to predict the degree of inhibition of ion uptake and relates survival time to a critical loss of ions from the internal fluid compartments. The prototype version of the IBM, developed to predict the toxicity of silver to fish, will also require further development and testing before it is ready for more general applicability.

Future Model Development Needs

The models reviewed herein will serve as useful tools to the analyst evaluating the fate and effects of metals in aquatic environments. However, a number of areas have been identified where further model development efforts are warranted. Figure 6-1 illustrates some of the important model components that should be included in a comprehensive aquatic exposure and risk assessment framework for metals. The fate and transport model itself is represented by the water column and sediment compartments shown at the center of the diagram. As shown to the left in Figure 6-1, a comprehensive fate and transport model framework will require the capability to perform dynamic simulations using temporally varying inputs. The time variable output that results will be suitable for use in a probabilistic analysis of exposure and effects.

A comprehensive fate and transport model should also simulate the basic parameters needed by an integrated geochemical submodel (also shown at the center of the diagram). The geochemical model should provide an improved representation of metal speciation and partitioning in comparison to the relatively simple methods incorporated in most models currently in use today. A sediment chemistry submodel (middle and lower right diagrams), capable of simulating the interactions of metals with sulfide and other binding phases, is also needed to provide a proper basis for evaluating interaction between the water column and sediment and to assess metal bioavailability and toxicity in sediments. Development of a BLM for sediments is one approach that would satisfy these needs.

The improved representation of chemical speciation that is called for above is consistent with the level of detail provided by the BLM (upper middle and upper right diagrams of Figure 6-1). The BLM is used to assess bioavailability and toxicity of water column metals as a function of site-specific and time-varying water quality. Finally, a bioaccumulation model, possibly a multicompartment pharmacokinetic model (middle right diagram), one that reflects metal essentiality, sequestration, and trophic-level transfer, is needed to complete a comprehensive exposure and risk assessment analysis. Each of these components is described in further detail in the remainder of this section.

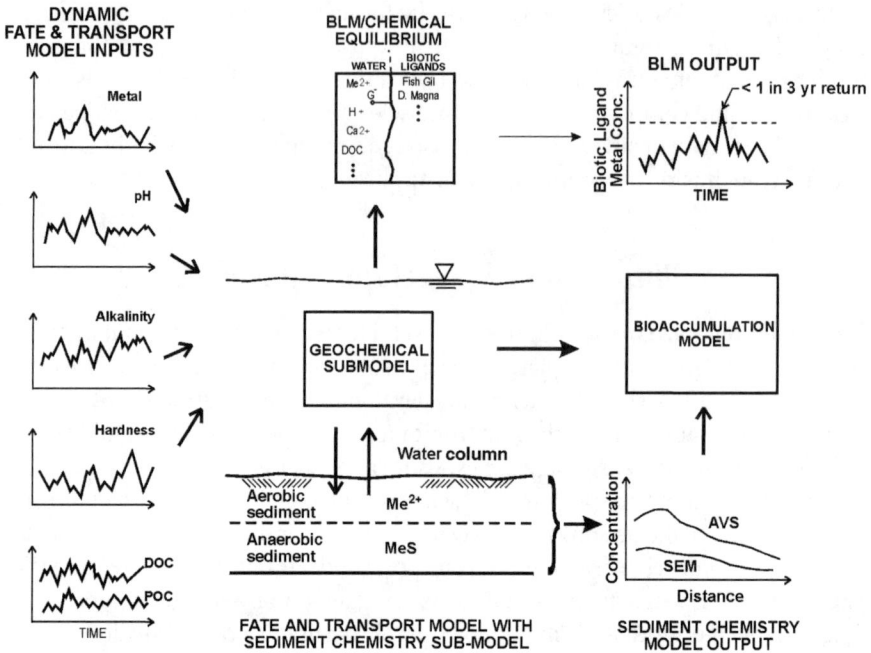

Figure 6-1 Components of a modeling framework for aquatic ecological risk assessments (ERAs) for metal (AVS = acid volatile sulfide, BLM = Biotic Ligand Model; SEM = simultaneously extracted metal)

Dynamic simulations and a probabilistic overlay

When steady-state analyses are made, regulatory agencies often require that critical low flow conditions and peak mass discharge rates be assumed. The concurrent application of multiple low probability events typically results in unnecessarily conservative results. Although in some instances this type of screening level analysis is sufficient to rule out the likelihood of a problem occurring, in other instances it can lead to an inefficient expenditure of resources. Thus, the capability to conduct probabilistic analyses and/or dynamic simulations, including Monte Carlo generation of inputs, will be needed if a meaningful characterization of exposure levels and effects is to be made. This capability is also needed in the regulatory setting of the U.S., where a method to evaluate the frequency of exceedance of the criteria is required in order to assess compliance. As discussed in Chapter 3, a number of models having these capabilities are currently available.

Chemical equilibrium model

A chemical equilibrium model suitable for characterizing the water column chemistry in a metals risk assessment should include inorganic speciation, complexation with DOM, adsorption on suspended particles, and solid-phase solubility constraints. Integration of a chemical equilibrium subroutine into fate and transport models will ultimately enhance their predictive capability. Pursuant to the recommendations made at the 1997 Pellston Conference (Kramer et al. 1997), the approach used in WHAM to characterize metal–DOM interactions has recently been incorporated in CHESS in conjunction with the BLM development effort. A refined procedure for simulating sorption of metals to particles is still needed. Investigations are under way that should provide a basis for this type of model refinement (e.g., Allen et al. 1998; Shi et al. 1998).

Situations may also exist where nonequilibrium conditions prevail, such as in the immediate vicinity of a discharge. When this occurs, the available chemical equilibrium models may not be suitable for use. As a result, a long-term goal of future research and model development efforts should be to improve the predictive capabilities of both geochemical and fate and transport models under nonequilibrium conditions.

Sediment chemistry model

Figure 6-1 also includes a sediment chemistry submodel, one that incorporates both an aerobic and an anaerobic sediment layer. A high-resolution version of a sediment chemistry model having this structure was described in Chapter 5 ("Modeling Metal Toxicity in Sediments," p 81). A comprehensive fate and transport model will need to incorporate this type of submodel in order to predict the exposure levels and bioavailability of metals in sediments and the potential for adverse effects. Recently, a simplified two-layer version of the simultaneously extracted metal and acid-volatile sulfide (SEM and AVS) model, one that includes both an aerobic surficial sediment layer and a deeper anaerobic layer, has been developed. This configuration is more amenable to incorporation in a fate and transport model for metals than a high-resolution vertical profile model because of its reduced computational requirements. The type of experimental and model development work needed to support the testing and refinement of this type of model should be continued. At the same time, the sediment model development effort should be extended to other metals. The simplified two-layer model also needs to be incorporated in a prototype metal fate and transport model, with field-scale testing ultimately required. This type of model will provide a way to simulate spatial distributions of AVS and SEM in surficial sediment layers (Figure 6-1, lower right diagram) and will facilitate the assessment of the likelihood of adverse aquatic impacts due to metals in sediments.

In situations where SEM exceeds AVS, a reliable model of porewater metal binding to both dissolved organic carbon (DOC) and particulate organic carbon (POC) will be needed. The approach to this type of model might be similar to the approach used for the water column, that is, it may be appropriate to develop a version of the BLM that is applicable to sediments.

Toxicity model

An acute toxicity model for metals should be able to account for the wide range in toxic effect levels that result from changes in water chemistry (including changes in pH, concentrations of DOM and suspended solids, concentrations of hardness cations, and availability of inorganic ligands such as carbonate ions). An example of such a model is the BLM, described in Chapter 5 (Di Toro et al. 1997, 2001; Paquin et al. 1998, 1999; Santore et al. 1998, 2001; Meyer et al. 1999; USEPA 1999a; Paquin, Gorsuch et al. 2002). The BLM has been developed using copper and silver toxicity data for freshwater fish and invertebrates. The BLM should be further validated using additional independent toxicity data sets for these same organisms, and the model parameter values refined as needed. It should also be extended in applicability to additional freshwater and saltwater organisms for copper and silver and to additional metals in general. Efforts in this regard have begun for cadmium, nickel, and zinc (e.g., de Schamphelaere, Heijerick et al. 2002; Santore et al. 2002). Further studies should be conducted to assess the applicability of this general approach to the evaluation of chronic toxicity and to the evaluation of sediment toxicity.

While the BLM development effort has been well received, it is not necessarily the only approach for predicting effects due to metals which may yield future benefits. It is also recommended that further effort be put forth with regard to the development of toxicity models that are based upon relationships between tissue residue levels and effects (e.g., CBR-type models) and to determine whether whole body or organ-specific tissue accumulation levels provide better predictive value. Models such as the IBM, which consider ionoregulatory effects, are also viewed as having potential applicability in the evaluation of effect levels of metals for endpoints other than survival. As elaborated upon in the next section, which discusses bioaccumulation modeling needs, an improved understanding of the potential for adverse effects via the dietary route of exposure is also called for.

Bioaccumulation model

Modeling studies of metal bioaccumulation have been limited and have largely focused on analysis of laboratory data for specific aquatic organisms. Further development of bioaccumulation models for metals, beyond the use of empirically derived bioconcentration and bioaccumulation factors, is clearly needed. Particular attention should be given to multicompartment pharmacokinetic models. This will lead to an improved understanding of how metals are sequestered in specific organs and how this sequestering affects depuration kinetics, toxicological effects on specific target organs, and bioavailability of metals to higher trophic-level organisms. It will also lead to an improved understanding of the importance of waterborne versus dietary routes of exposure. This effort will require a combination of laboratory, field, and modeling studies to further test model formulations for metal bioaccumulation and to determine appropriate model parameters (e.g., internal transfer rates, metal partitioning in specific organs, metallothionein induction rates).

Significant differences in metal bioaccumulation are expected for different trophic levels and among species of the same trophic level. Reasons for this are not well understood at this time, and further research is needed. Similarly, an improved understanding of the processes that control the active uptake and elimination of essential metals and how best to represent these processes in bioaccumulation models is also needed. Incorporation of these refinements in existing bioaccumulation models will ultimately lead to an improved ability to model metal bioaccumulation both across species at a given trophic level and across trophic levels in more complex aquatic food webs.

Development of a way to predict the bioaccumulation of metals from sediments, and more importantly, to advance the understanding of how the level of accumulation may be related to chronic effects, is a remaining area that warrants further research in coming years. Also, because metals often coexist with other metals and HOCs in the environment, a long-term goal in bioaccumulation modeling should be the assessment of synergistic effects among contaminants. For example, George (1989) reports that cytochrome P-450–dependent ethoxyresorufin-O-dealkylase (EROD) activity, which is important in the metabolism and detoxification of polyaromatic hydrocarbons (PAHs), was strongly inhibited in fish by cadmium exposure.

Concluding Remarks

The recommendations in this concluding section are in many ways consistent with the metals-related research needs identified at the 1997 Pellston Workshop on metals WQC (Bergman and Dorward-King 1997). With regard to fate and transport modeling of metals (Schnoor et al. 1997), the Pellston recommendations were that, over the short-term, a probabilistic modeling approach be adopted for riverine systems, rather than relying on a simple static dilution model, and that models be developed to protect both the water column and sediment. It was also recommended that discharge permits be developed on a watershed scale. Longer-term recommendations identified the need for a time variable modeling approach suitable for multiple loads and multiple routes of chemical exposure to biota, including the need for use of a food chain model. Short-term and long-term model testing was recommended. Modeling research priorities that were identified included

    1) testing of hypotheses that link metal exposure to effects at the target organ (i.e., testing of the utility of the biotic ligand–type modeling approach) and incorporating the results into model algorithms,

2) incorporating chemical speciation into fate and transport model frameworks, and

3) developing models that include sorption and desorption kinetics and speciation.

Process and experimental research priorities in support of model development focused on sediment–water and water–particle exchange processes, including mechanisms and rates of metal oxidation and release from in-place and resuspended sediments.

Although significant progress has been made in many of the preceding areas in the short time since the 1997 Pellston meeting, much remains to be accomplished. This is demonstrated by the fact that many of these same matters were renewed topics of discussion at a May 2003 Pellston Workshop (Pensacola, Florida, USA) on the hazard identification approach for metals. It is expected that steady progress will continue to be made as the scientific community responds to the expressed needs of both government and industry.

APPENDIX
Partial List of Sources of Available Models

Note: Prospective users should inquire about updated versions of the models reviewed in this report.

Fate and Transport Models

CTAP
Computer Resources
HydroQual, Inc.
1200 MacArthur Blvd
Mahwah NJ, USA 07430
Phone: (201) 529-5151

DELFT3D
WL|Delft Hydraulics
P.O. Box 177
2600 MH Delft
The Netherlands
Phone: +31 (0) 15 2858585
www.wldelft.nl/soft/d3d

DJOC
O'Connor DJ. 1988. Models of sorptive toxic substances in freshwater systems. I: Basic equations. *J Environ Engin* 114(3). ASCE, ISSN Paper Nr 22485.

O'Connor DJ. 1988. Models of sorptive toxic substances in freshwater systems. II: Lakes and Reservoirs. *J Environ Engin* 114(3). ASCE, ISSN Paper Nr 22486.

O'Connor DJ. 1988. Models of sorptive toxic substances in freshwater systems. III: Streams and Rivers. *J Environ Engin* 114(3). ASCE, ISSN Paper Nr 22487.

EXAMS II ver. 2.98.04
USEPA Center for Exposure Assessment Modeling (CEAM)
Ecosystems Research Division
National Exposure Research Laboratory
960 College Station Road
Athens GA, USA 30605-2700
Phone: (706) 355-8400
Fax: (706) 355-8302
www.epa.gov/ceampubl/swater/index.htm
Email: ceam@epamail.epa.gov

HSPF
USEPA Center for Exposure Assessment Modeling (CEAM)
Ecosystems Research Division
National Exposure Research Laboratory
960 College Station Road
Athens GA, USA 30605-2700
Phone: (706) 355-8400
Fax: (706) 355-8302
www.epa.gov/ceampubl/swater/index.htm
Email: ceam@epamail.epa.gov

QWASI
Mackay D. 1991. Multimedia environmental models. Chelsea MI, USA: Lewis Publishers (This book includes a diskette with programs.)

RECOVERY

U.S. Army Corps of Engineers
U.S. Army Engineers Research & Development Center
3909 Halls Ferry Road
Vicksburg MS, USA 39180-6199
Phone: (601) 634-3784
Fax: (601) 634-3129
www.wes.army.mil/el/elmodels/

RIVRISK (EPRI)

Electric Power Software Center
Phone: (800) 313-3774
www.epri.com

SLSA

Computer Resources
HydroQual, Inc.
1200 MacArthur Blvd
Mahwah NJ, USA 07430
Phone: (201) 529-5151

SMPTOX3

USEPA Center for Exposure Assessment Modeling (CEAM)
Ecosystems Research Division
National Exposure Research Laboratory
960 College Station Road
Athens GA, USA 30605-2700
Phone: (706) 355-8400
Fax: (706) 355-8302
www.epa.gov/ceampubl/swater/index.htm
Email: ceam@epamail.epa.gov

USES

National Institute of Public Health and the Environment (RIVM)
Centre for Substances and Risk Assessment (CSR)
P.O. Box 1
3720 BA Bilthoven
The Netherlands
Phone: +31 30 274 30 04
Fax: +31 30 274 44 01
http://arch.rivm.nl/csr/risk.html

EUSES

European Chemicals Bureau
JRC Environmental Institute
I-21020 Ispra (Varese)
Italy
Phone: +390 332 78 58 66
Fax: +390 332 78 58 62
Email: euses.euses@jrc.it

WASP5

(Note: WASP6, including a graphical user interface, is now available.)

USEPA Center for Exposure Assessment Modeling (CEAM)
Ecosystems Research Division
National Exposure Research Laboratory
960 College Station Road
Athens GA, USA 30605-2700
Phone: (706) 355-8400
Fax: (706) 355-8302
www.epa.gov/ceampubl/swater/
ceam@epamail.epa.gov

WASTOX

Computer Resources
HydroQual, Inc.
1200 MacArthur Blvd
Mahwah NJ, USA 07430
Phone: (201) 529-5151

WQAM

Mills WB, Dean JD, Porcella DB, Gherini SA, Hudson RJM, Frick WE, Rupp GL, Bowie GL. 1982. Water quality assessment: A screening procedure for toxic and conventional pollutants. Part 1, 2. and 3. EPA-600/6-82-004a, b, and c.

Mills WB, Porcella DB, Ungs MJ, Gherini SA, Summers KV, Lingfung Mok, Rupp GL, Bowie GL, Haith DA. 1985. Water quality assessment: A screening procedure for toxic and conventional pollutants in surface and ground water. Part 1. EPA-600-6-85-002a.

Chemical Equilibrium Models

CHESS

Computer Resources
HydroQual, Inc.
1200 MacArthur Blvd
Mahwah NJ, USA 07430
Phone: (201) 529-5151

MINEQL+

Environmental Research Software
16 Middle Street
Hallowell ME, USA 04347
Phone: (207) 622-3340
Fax: (207) 622-3340
www.mineql.com/mineql.html

MINTEQA2

USEPA Center for Exposure Assessment Modeling (CEAM)
Ecosystems Research Division
Natl Exposure Research Laboratory
960 College Station Road
Athens GA, USA 30605-2700
Phone: (706) 355-8400
Fax: (706) 355-8302
www.epa.gov/ceampubl/mmedia/
Email: ceam@epamail.epa.gov

WHAM

The Center of Ecology and Hydrology
CEH Windermere
Institute of Freshwater Ecology
The Ferry House
Far Sawrey
Ambleside
Cumbria, LA22 OLP, UK
Phone: (+44) 15394 42468
Fax: (+44) 15394 46914
http://windermere.ceh.ac.uk/

Bioaccumulation Models

AQUATOX

USEPA
Office of Water (4101M)
1200 Pennsylvania Ave., NW
Washington DC, USA 20460
Telephone: 202-260-9821
Email: ow_general@epa.gov
www.epa.gov/waterscience/models/aquatox/

EPA-823-C-00-001: AQUATOX a modular fate & effects model for aquatic ecosystems-CDROM.*

EPA-823-R-00-006: Volume 1 User's Manual.*

EPA-823-R-00-007: Volume 2 Technical Documentation.*

EPA-823-R-00-008: Volume 3 Model Validation Reports.*

Thomann Model

Manhattan College
Department of Environmental Engineering
Manhattan College Parkway
Riverdale NY, USA 10471
Phone: (718) 862-7276

Toxicity Models

BLM

Computer Resources
HydroQual, Inc.
1200 MacArthur Blvd
Mahwah NJ, USA 07430
Phone: (201) 529-5151

*Can be obtained from National Service Center for Environmental Publications (NSCEP), which may be reached by calling (800) 490-9198 or on the Internet, www.epa.gov or www.epa.gov/ncepihom.

Abbreviations

2-H	two-dimensional horizontal plane
2-V	two-dimensional vertical plane
AESOP	advanced ecological systems operating program
AQUATOX	Aquatic Toxicity model
AS	analytical solution
ATPase	adenosine triphosphatase
AVS	acid-volatile sulfide
BASS	Bioaccumulation and Aquatic System Simulator
BAF	bioaccumulation factor
BCF	bioconcentration factor
BIOTRANSPEC	a further adaptation of the TRANSport and SPECiation model
BLM	Biotic Ligand Model
BSAF	biota–sediment accumulation factor
CAUC	critical area under the curve
CBR	critical body residue
CCOD	current contents on diskette
CH3D	U.S. Army Corps of Engineers Chesapeake Bay model
CHAOS	Complexation by Humic Acids in Organic Soils
CHESS	Chemical Equilibrium in Soils and Solutions model
CHNTRN	Channel Transport model
CM	complete mixing volume
CMA	Chemical Manufacturers Association
CMV	completely mixed volume
CTAP	Chemical Transport and Analysis Program
DELFT3D	Delft 3D model
DDE	dicholorodiphenyldichloroethylene
DHI	Danish Hydraulic Institute
DJOC	Donald J. O'Connor model
DOC	dissolved organic carbon
DOM	dissolved organic matter
DYNHYD5	a link-node hydrodynamic model applicable to predominantly 1-D settings
DYNTOX	Dynamic Toxics model

EEC	estimated environmental concentration
EC	European Commission
EC50	effective concentration to 50% of a test population
ECOM	Estuary, Coastal, Ocean Model
ECOMSED	Estuary, Coastal, Ocean Model-Sediment
EFDC	Environmental Fluid Dynamics Code
EPRI	Electric Power Research Institute
EqP	equilibrium partitioning
ERA	ecological risk assessment
EROD	ethoxyresorufin-O-deethylase
ESG	equilibrium partitioning sediment guideline
EUSES	European Union System for the Evaluation of Substances
EUTRO	kinetic subroutine for use in eutrophication problems for the fate and transport model WASP5
EXAMS	Exposure Analysis Modeling System
FETRA	Sediment/Radionuclide Transport model
FGETS	Food and Gill Exchange of Toxic Substances model
FIAM	Free-Ion Activity Model
FIFRA	Federal Insecticide, Fungicide, and Rodenticide Act
FORTRAN	Formula Translator
GIT	gastrointestinal tract
GMIII	General Motors III model
GMIN	model utilizing the free energy minimization approach, which relies on sophisticated numerical methods to improve convergence and to avoid local minima
HMW	high molecular weight
HOC	hydrophobic organic chemical
HSP	Hydrocomp Simulation Program
HSPF	Hydrologic Simulation Program-FORTRAN
IBM	ion balance model
ILL	incipient lethal level
IMW	intermediate molecular weight

LA50	lethal accumulation level resulting in 50% mortality
LC50	lethal concentration for 50% of test organisms
LMW	low molecular weight
MCM	Mercury Cycling Model
META	Metal Exposure and Transformation Assessment
META4	WASP-based Metal Exposure and Transformation Assessment model
MEXAMS	Metals Exposure Analysis Modeling System
MICHRIV	Michigan River model
MICROQL	a less comprehensive version of MINEQL, developed for classroom microcomputer use
MIKE21	a 3-D model that numerically solves the controlling hydrodynamic and fate and transport equations for a range of water quality variables
MINEQL	Minicomputer Equilibrium model
MINEQL+	Minicomputer Water Equilibrium model +
MINTEQ	Minicomputer Water Equilibrium model
MT	metallothionein
NPDES	National Pollutant Discharge Eliminations System
NICA	Nonideal Competitive Adsorption model
NOAA	National Oceanographic and Atmospheric Administration (U.S.)
NOEC	no-observed-effect concentration
PAH	polyaromatic hydrocarbon
PAWTOXIC	Pawtuxent Toxics
PB-PK	physiologically based pharmacokinetic
PCB	polychlorinated biphenyl
PCDD	polychlorinated dibenzo-p-dioxin
PCDF	polychlorinated dibenzofuran
PDM	Probabilistic Dilution Model
PEC	predicted environmental concentration
PHREEQE	one of several models developed by the U.S. Geological Survey (USGS) from an ion association model for seawater
PNEC	predicted no-effect concentration
POC	particulate organic carbon
PTR	Pesticide Transport and Runoff model

QUAL-II	1-D steady-state stream water quality model
QWASI	Quantitative Water–Air Sediment Interaction
RAND	equilibrium model that evolved from early computer codes
RCATOX	Row-Column AESOP for Toxics model
REDEQL	equilibrium model that evolved from early computer codes
RIVEQL	River Quality model
RIVM	National Institute of Public Health and the Environment (The Netherlands)
RIVRISK	River Risk model
SAB	Science Advisory Board (U.S.)
SEM	simultaneously extracted metal
SERATRA	Sediment Contaminant, i.e., Radionuclide Transport Model
SETAC	Society of Environmental Toxicology and Chemistry
SLSA	Simplified Lake and Stream Analysis
SMPTOX	Simplified Method Program–Variable-Complexity Stream Toxics model
SOLMNEQ	one of several models developed by the U.S. Geological Survey (USGS) from an ion association model for seawater
SSNS	Steady-state numberical solution
SWM	Stanford Watershed Model
SYVAC	Systems Variability Analysis Code
TBTO	tributyl-tin oxide
TDS	total dissolved solid
TGD	technical guidance document
TMDL	total maximum daily load
TOC	total organic carbon
TOXI4	chemical fate model created when WASP4 is linked with the chemical fate subroutine
TOXI5	kinetic subroutine distributed with WASP5 for simulating chemical fate and transport
TOXIWASP	the original version of WASP incorporating a chemical fate and transport subroutine
TRANSPEC	TRANSport and SPECiation model
TSS	total suspended solid
TVNS	time-variable numerical solution

USEPA	U.S. Environmental Protection Agency
USES	Uniform System for the Evaluation of Substances
USGS	U.S. Geological Survey
WASP	Water Quality Analysis Simulation Program
WASP5	Water Quality Analysis Simulation Program, Version 5
WASTOX	Water Quality Analysis Simulation of Toxics Model
WATEQ	one of several models developed by the U.S. Geological Survey (USGS) from an ion association model for seawater
WHAM	Windermere Humic Aqueous Model
WQAM	Water Quality Assessment Methodology
WQC	water quality criterion
ZID	zone of initial dilution

References

Adams WJ, Kimerle RA, Mosher RG. 1985. Aquatic safety assessment of chemicals sorbed to sediments. In: Cardwell RD, Purdy R, Bahner RC, editors. Aquatic toxicology and hazard assessment: 7th symposium. Philadelphia PA, USA: ASTM. STP 854. p 429-453.

Allen HE, Hansen DJ. 1996. The importance of trace metal speciation to water quality criteria. *Water Environ Res* 68(1):42-54.

Allen HE, Zong Y, Lu Y. 1998. Copper partitioning between river water and suspended solids. Newark DE, USA: University of Delaware, Department of Civil Engineering prepared for USEPA, HydroQual, Inc. 74 p.

Allison JD, Brown DS, Novo-Gradac KJ. 1991. MINTEQA2/PRODEFA2, A geochemcal assessment model for environmental systems. Version 3.0, User's manual. Athens GA, USA: USEPA, ERL, ORD. EPA-600-3-91-021.

Allison JD, Perdue EM. 1994. Modeling metal-humic interactions with MINTEQ2. In: Senesi N, Miano TM, editors. Humic substances in the global environment and implications on human health. Amsterdam, NL: Elsevier Science. p 927-942.

Ambrose RB. 1986. WASP3, a hydrodynamic and water quality model-model theory, user's manual, and programmers guide. Athens GA, USA: USEPA, ERL. EPA-600-3-86-034.

Ambrose RB. 1988. WASP4, a hydrodynamic and water quality model-model theory, user's manual and programmers guide. Athens GA, USA: USEPA, ERL. EPA-600-3-87-039.

Ambrose RB, Hill SI, Mulkey LA. 1983. User's manual for the chemical transport and fate model (TOXIWASP). Version 1. Athens GA, USA: USEPA, ERL. EPA-600-3-83-005.

Ambrose RB, Wool TA, Martin JL. 1993. The water quality analysis simulation program, WASP5. Version 5.10, Part A: Model documentation. Athens GA, USA: USEPA, ORD, ERL.

Ankley GT, Berry WJ, De Rosa LD, Di Toro DM, Hansen DJ, Reiley M. 1994. Briefing report to the EPA Science advisory board on the equilibrium partitioning approach to predicting metal bioavailability in sediments and the derivation of sediment quality criteria for metals. Washington DC, USA: USEPA, OW and ORD. EPA-822-D-94-002.

Ankley GT, Berry WJ, Di Toro DM, Hansen DJ, Hoke RA, Mount DR, Reiley MC, Schwartz RC, Zarba CS. 1996. Use of equilibrium partitioning to establish sediment quality criteria for nonionic chemical: A reply to Iannuzzi et al. letter to the editor. *Environ Toxicol Chem* 15(7):1019-1024.

Ankley GT, Di Toro DM, Hansen WJ, Berry WJ. 1996. Technical basis and proposal for deriving sediment quality criteria for metals. *Environ Toxicol Chem* 15(12):2056-2066.

Ankley GT, Mattson V, Leonard E, West C, Bennett J. 1993. Predicting the acute toxicity of copper in freshwater sediments: Evaluation of the role of acid volatile sulfide. *Environ Toxicol Chem* 12(2):315-320.

Backes CA, Tipping E. 1987. Aluminum complexation be an aquatic humic fraction under acidic condition. *Water Res* 21(2):211-216.

Ball JW, Jenne EA, Cantrell MW. 1981. WATEQ3: A geochemical model with uranium added. Washington DC: U.S. Geological Survey. Open File Report 81-1183.

Ball JW, Nordstrom DK, Zachmann DW. 1987. WATEQ4F: A personal computer FORTRAN translation of the geochemical model WATEQ2 with revised data base. Washington DC: U.S. Geological Survey. Open File Report 87-50. 108 p.

Barber MX, Suarez LA, Lassiter RR. 1991. Modelling bioaccumulation of organic pollutants in fish with an application to PCBs in Lake Ontario salmonids. *Can J Fish Aquat Sci* 48:318-337.

Barron MG, Stehly GR, Hayton WL. 1990. Pharmacokinetic modeling in aquatic animals I. Models and concepts. *Aquat Toxicol* 18:61-86.

Bassett RL, Melchior DC. 1990. Chemical modeling of aqueous systems. In: Melchior DC, Bassett RL, editors. Chemical modeling of aqueous systems II. Washington DC, USA: ACS. 556 p.

Baumgartner DJ, Frick WE, Roberts PJW. 1994. Dilution models for effluent discharges. 3rd ed. Newport OR, USA: USEPA, Pacific Oceans Systems Branch. EPA-600-R-94-086.

Benedetti MF, Milne CJ, Kinniburgh DG, Van Riemsdijk, Koopal LK. 1995. Metal ion binding to humic substances: Application of the non-ideal competitive adsorption model. *Environ Sci Technol* 29(2):446-457.

Benson WH, Birge WJ. 1985. Heavy metal tolerance and metallothionein induction in fathead minnows: results from field and laboratory investigations. *Environ Toxicol Chem* 4:209-217.

Berding V, Schwartz S, Matthies M. 1999. Visualisation of the complexities of EUSES. *Environ Sci Pollut Res* 6(1):37-43.

Bergman HL, Dorward-King EJ, editors. 1997. Reassessment of metals criteria for aquatic life protection: Priorities for research and implementation. Proceedings of the Pellston Workshop on Reassessment of Metals Criteria for Aquatic Life Protection; 1996 Feb 10-14; Pensacola, FL, USA. Pensacola FL, USA: SETAC. 114 p.

Berry WJ, Cantwell MG, Edwards PA, Serbst JR, Hansen DJ. 1998. Predicting the toxicity of sediments spiked with silver. *Environ Toxicol Chem* 18(1):40-48.

Berry WJ, Hansen DJ, Mahony LD, Robson DL, Di Toro DM, Shipley BP, Rogers B, Corbin JM, Boothman WS. 1996. Predicting the toxicity of metal-spiked laboratory sediments using acid-volatile sulfide and interstitial water normalizations. *Environ Toxicol Chem* 15(12):2067-2079.

Bhavsar SP, Diamond ML, Evans LJ, Gandhi N, Nilsen J, Cypas P. 2003. Development of a coupled metal speciation-fate model for lakes: TRANSPEC. *Environ Toxicol Chem*. Forthcoming.

Bicknell BR, Imhoff JC, Kittle JL, Donigian AS, Johanson AC. 1993. Hydrologic simulation program-FORTRAN (HSPF): User's manual for release 10.0. Athens GA, USA: USEPA, ERL. EPA-600-3-84-066.

Bills T, Klaine SJ, LaPoint TW, Cobb G, Forsythe B, Wenholz M. 1997. Influence of water quality parameters on acute silver nitrate toxicity to *Daphnia magna*. Final Report Nr TIWET-9506.1. Submitted to the Silver Council c/o the National Association of Photographic Manufacturers by the Department of Environmental Toxicology, The Institute of Wildlife and Environmental Toxicology (TIWET), Clemson University, Pendleton SC, USA.

Bird GA, Stephensen M, Cornett RJ. 1993. The surface water model for assessing Canada's nuclear fuel waste disposal concept. *Waste Manage* 13:153-170.

Black MC, McCarthy JF. 1988. Dissolved organic macromolecules reduce the uptake of hydrophobic organic contaminants by the gills of rainbow trout *(Salmo gairdneri)*. *Environ Toxicol Chem* 7(7):593-600.

Blumberg AF, Khan LA, St. John JP. 1999. Three-dimensional hydrodynamic model of New York Harbor region. *J Hydraulic Engineering ASCE* 125(8):799-816.

Blumberg AF, Mellor GL. 1987. A description of a three-dimensional coastal ocean circulation model. In: Heaps N, editor. Three-dimensional coastal ocean models. Washington DC, USA: American Geophysical Union. p 1-16.

Blumberg AF, Signell RP, Jenter HL. 1993. Modeling transport processes in the coastal ocean. *J Mar Environ Engineering* 1:3-52.

Boyer JM, Chapra SC, Ruiz CE, Dortch MS. 1994. Recovery: A mathematical model to predict the temporal response of surface water contaminated sediments. Vicksburg MS, USA: U.S. Army Engineer Experiment Station. Technical Report W-94-4.

Breck JE. 1988. Relationship among models for acute toxic effects: applications to fluctuating concentrations. *Environ Toxicol Chem* 7(9):775-778.

Brown DA, Parsons TR. 1978. Relationship between cytoplasmic distribution of mercury and toxic effects to zooplankton and chum salmon (*Oncorhynchus keta*) exposed to mercury in a controlled ecosystem. *J Fish Res Board Can* 35:880-884.

Brown LC, Barnwell TO. August 1985. Computer program documentation for the Enhanced Stream Water Quality Model QUAL2E. Prepared for the USEPA Environmental Research Laboratory. Athens, Georgia. EPA/600/3-85/065.

Burkhard LP. 1998. Comparison of two models for predicting bioaccumulation of hydrophobic organic chemicals in a Great Lakes food web. *Environ Toxicol Chem* 17(3):383-393.

Burns LA. 1983. Fate of chemicals in aquatic systems: Process models and computer codes. In: Swann RL, Eschenroeder A, editors. Fate of chemicals in the environment, compartmental and multimedia models for predictions. Washington DC, USA: ACS. Symposium Series 225. p 25-40.

Burns LA. 1990. Exposure analysis modeling system: Users guide for EXAMS II, Version 2.94. Athens GA, USA: USEPA, Research Laboratory. EPA-600-3-89-084.

Burns LA. 1997. Exposure analysis modeling system (EXAMS II), User's guide for Version 2.97. Athens GA, USA: USEPA, Research Laboratory. 133 p.

Burns LA, Cline DM. 1985. Exposure analysis modeling system reference manual for EXAMS II. Athens GA, USA: USEPA, ERL. EPA-600-3-85-038.

Burns LA, Cline DM, Lassiter RR. 1982. Exposure analysis modeling system (EXAMS): User manual and system documentation. Athens GA, USA: USEPA, Research Laboratory. EPA-600-3-82-023, PB82-258096.

Bury NR, Galvez F, Wood CM. 1999. Effects of chloride, calcium and dissolved organic carbon on silver toxicity: comparison between rainbow trout and fathead minnows. *Environ Toxicol Chem* 18(1):56-62.

Campbell PGC. 1995. Interactions between trace metals and aquatic organisms: A critique of the free-ion activity model. In: Tessier A, Turner DR, editors. Metal speciation and bioavailability in aquatic systems. New York NY, USA: John Wiley and Sons, Inc. p 45-102.

Carbonaro RF. 1999. The experimental calibration of a sediment metal flux model [MSc thesis]. Bronx NY, USA: Manhattan College. Dominic Di Toro, advisor.

[CDNR/CDPHE] Colorado Department of Natural Resources and Colorado Department of Public Health and Environment. 1996. Use attainability assessment Alamosa River watershed through 1996. Denver CO, USA: CDPHE. Draft Report Prepared by Posey HH, Woodling J, Campbell A, Pendleton JA. 246 p.

Chang SI, Reinfelder JR. 2000. Bioaccumulation, subcellular distribution, and trophic transfer of copper in a coastal marine diatom. *Environ Sci Technol* 34:4931-4935.

Chapman BM. 1982. Numerical simulation of the transport and speciation of nonconservative chemical reactants in rivers. *Water Resour Res* 18(1):155-167.

Chapman PM, Allen HE, Godtfredsen K, Z'Graggen M. 1996. Evaluation of bioaccumulation factors in regulating metals. *Environ Sci Technol* 30(10):448A-452A.

Chapra SC. 1982. Long-term models of interaction between solids and contaminants in lakes [D Phil dissertation]. Ann Arbor MI, USA: The University of Michigan.

Chapra SC. 1997. Surface water-quality modeling. McGraw-Hill series in water resources and environmental engineering. New York NY, USA: McGraw-Hill. 844 p.

Chapra SC, Reckhow KH. 1983. Engineering Approaches for Lake Management. Volume 2: Mechanistic modeling. Butterworth Woburn MA, USA: Ann Arbor Science.

Chen KY. 1996. Modeling the fate of copper discharged to San Francisco Bay. *J Environ Engineering* 122(10):924-934.

Connolly JP. 1985. Predicting single-species toxicity in natural water systems. *Environ Toxicol Chem* 4:573-582.

Connolly JP. 1991. Application of a food chain model to polychlorinated biphenyl contamination of the lobster and winter flounder food chains in New Bedford Harbor. *Environ Sci Technol* 25(4):760-770.

Connolly JP, Thomann RV. 1985. WASTOX: A framework for modeling the fate of toxic chemicals in aquatic environments. Part 2: Food chain. Gulf Breeze FL and Duluth MN, USA: USEPA. 33 p.

Connolly JP, Thomann RV. 1992. Modeling the accumulation of organic chemicals in aquatic food chains. In: Schnoor JL, editor. Fate of pesticides and chemicals in the environment. New York NY, USA: John Wiley and Sons, Inc. p 385-406.

Connolly JP, Winfield RP, 1984. WASTOX: A framework for modeling toxic chemicals in aquatic systems. Part 1: Exposure concentration. Gulf Breeze FL, USA: USEPA. EPA-600-3-84-077.

Coombs TL, George SG. 1978. Mechanisms of immobilization and detoxication of metals in marine organisms. In: McLusky DS, Berry JA, editors. Physiology and behaviour of marine organisms. Oxford, UK: Pergamon Press. p 179-187.

Cowan CE, Versteeg DJ, Larson RJ, Kloepper-Sams PJ. 1995. Integrated approach for environmental assessment of new and existing substances. *Regul Toxicol Pharmacol* 21:3-31.

Davies PH. 1997. Acute and chronic toxicity of silver to aquatic life at different water hardnesses and effects of mountain and plains sediments on the bioavailability and toxicity of silver. Fort Collins CO, USA: Colorado Division of Wildlife.

Davies PH, Goettl Jr JP, Sinley JR. 1978. Toxicity of silver to rainbow trout (*Salmo gairdneri*). *Water Res* 12:113-117.

Delft Hydraulics. 1998. DELFT3D software specifications. Delft, NL: Delft Hydraulics Institute. 20 p.

Delos CG, Richardson WL, DePinto JV, Rodgers DW, Rygwelski K, Wethington, Ambrose RB, St. John JP. 1984. Technical guidance manual for performing waste load allocations. Book II: Streams and rivers. Washington DC, USA: USEPA, OW Regulations and Standards, Monitoring and Data Support Division. EPA-440-4-84-022.

de Rooij NM, Smallegange RAJ, Smits JGC, Temminghoff EJM, Plette ACC, Bril J. 1999. Methodology for determination of heavy metal standards for soil, phase 2: Development of models and measuring techniques. Delft, NL: National Institute for Public Health and the Environment (RIVM-CSR), prepared by AB-DLO, WAU-SSPN and WL Delft Hydraulics.

De Schamphelaere KAC, Heijerick DG, Janssen CR. 2002. Refinement and field validation of a biotic ligand model predicting acute copper toxicity to *Daphnia magna*. Special Issue: The Biotic Ligand Model for Metals–Current Research, Future Directions, Regulatory Implications. *Comp Biochem Physiol* Part C 133(1-2):243-258.

De Schamphelaere KAC, Janssen CR. 2002. A biotic ligand model predicting acute copper toxicity for *Daphnia magna*: The effects of calcium, magnesium, sodium, potassium and pH. *Environ Sci Technol* 36(1):48-54.

Diamond JM, Gerardi C, Leppo E, Miorelli T. 1997. Using a water-effect ratio approach to establish effects of an effluent-influenced stream on copper toxicity to the fathead minnow. *Environ Toxicol Chem* 16(7):1480-1487.

Diamond ML, Mackay D, Cornett RJ, Chant LA. 1990. A model of the exchange of inorganic chemicals between water and sediments. *Environ Sci Tech* 24(5):713-722.

Diamond ML, Mackay D, Welbourn PM. 1992. Models of multi-media partitioning of multi-species chemicals: the fugacity/aquivalence approach. *Chemosphere* 25:1907-1921.

Dilks DW, Helfand JS, Bierman Jr VJ. 1994. Sediment quality modeling in response to proposed sediment quality criteria. Proceedings WEFTEC 67th Annual Conference and Exposition, Volume 4, Surface Water Quality and Ecology; 1994 Oct 15-19; Chicago IL, USA. Alexandria VA, USA: Water Environment Federation. p 707-713. 713 p.

Dilks DW, Hjelfand JS, Bierman Jr VJ. 1995. Development and application of models to determine sediment quality criteria-driven permit limits for metals. Proceedings of Toxic Substances in Water Environments: Assessment and Control sponsored by WPCA of Ohio and SETAC; 1995 May 14-17; Cincinnati, OH, USA. Alexandria VA, USA: Water Environment Federation. p 1-37 – 1-48.

Di Toro DM. 1984. Probability model of stream quality due to runoff. *J Environ Engineering ASCE* 110(3):607-628.

Di Toro DM. 1987. Modeling the fate of toxic chemicals in surface waters. Short course notes from 8th Annual SETAC Meeting; 1987 Nov; Pensacola FL, USA.

Di Toro DM. 2001. Cadmium and iron. In: Sediment flux modeling. New York NY, USA: John Wiley and Sons, Inc.

Di Toro DM, Allen HE, Bergman HL, Meyer JS, Paquin PR, Santore RC. 2001. A biotic ligand model of the acute toxicity of metals. I. Technical basis. *Environ Toxicol Chem* 20(10):2383-2396.

Di Toro DM, Fitzpatrick JJ. 1993. Chesapeake Bay sediment flux model. Vicksburg MS, USA: U.S. Army Corps of Engineers, Waterways Experiment Station, Environmental Laboratory. 200 p.

Di Toro DM, Fitzpatrick JJ, Thomann RV. 1981. Water quality analysis simulation program (WASP) and model verification program (MVP). Grosse Ile MI, USA: Hydroscience Inc. for USEPA, Grosse Ile Laboratory, NERC. Revised 1983.

Di Toro DM, Hansen DJ, McGrath JA, Berry WJ. 1999. Predicting the toxicity of metals in sediments. In: Allen HE, Bell HE, Berry WJ, Di Toro DM, Hansen DJ, Meyer JS, Mitchell JL, Paquin PR, Reiley MC, Santore RC. Integrated approach to assessing the bioavailability and toxicity of metals in surface waters and sediments. Washington DC, USA: USEPA, Science Advisory Board, OW, ORD. EPA-822-E-99-001. p 2.22–2.37.

Di Toro DM, Mahony JD, Gonzalez AM. 1996. Particle oxidation model of synthetic FeS and sediment acid volatile sulfide. *Environ Toxicol Chem* 15(12):2156 2167.

Di Toro DM, Mahony JD, Hansen DJ, Berry WJ. 1996. A model of the oxidation of iron and cadmium sulfide in sediments. *Environ Toxicol Chem* 15(12):2168-2186.

Di Toro DM, Mahony JD, Hansen DJ, Scott KJ, Carlson AR, Ankley GT. 1992. Acid volatile sulfide predicts the acute toxicity of cadmium and nickel in sediments. *Environ Sci Technol* 26(1):96-101.

Di Toro DM, Mahony JD, Hansen DJ, Scott KJ, Hicks MB, Mayr SM, Redmond MS. 1990. Toxicity of cadmium in sediments: The role of acid volatile sulfide. *Environ Toxicol Chem* 9:1487-1502.

Di Toro DM, McGrath JA, Hansen DJ, Berry WJ. September 2003. Predicting the acute and chronic toxicity of metals in sediments using organic carbon normalized SEM and AVS: A sediment biotic ligand model. In preparation.

Di Toro DM, O'Connor DJ, Thomann RV, St. John JP. 1981. Analysis of the fate of chemicals in receiving waters, Phase 1. Washington DC, USA: CMA, Aquatic Research Task Group

Di Toro DM, Paquin PR. 1984. Time variable model of the fate of DDE and lindane in a quarry. *Environ Toxicol Chem* 3:335-353.

Di Toro DM, Santore RC, Paquin PR. 1997. Chemistry of copper bioavailability. I: Model of acute copper toxicity to fish, 18th SETAC Annual Meeting Abstract Book; Nov 1997; San Francisco, CA, USA. Pensacola FL, USA: SETAC.

Di Toro DM, Zarba CS, Hansen DJ, Berry WJ, Swartz RC, Cowen CE, Pavlou SP, Allen HE, Thomas NA, Paquin PR. 1991. Technical basis for establishing sediment quality criteria for non-ionic organic chemicals using equilibrium partitioning. *Environ Toxicol Chem* 10 (12):1541-1583.

Donigian AS, Imhoff JC, Bicknell BR, Kittle JL. 1984. Application guide for Hydrological Simulation Program - Fortran (HSPF). Athens GA, USA: USEPA, Environmental Research Laboratory. EPA-600/3-84-065.

Donigian AS, Patwardhan AS. 1992. Modeling nutrient loadings from croplands in the Chesapeake Bay watershed. In: Karamouz M, editor. Proceedings of water resources. Water Forum; 1992 Aug 2-6; Baltimore, MD, USA. New York NY, USA: ASCE Env Engr Div. p 817-822.

Dunbar LE. 1996. Derivation of a site-specific dissolved copper criteria for selected freshwater streams in Connecticut. Falmouth MA, USA: Connecticut DEP, Water Toxics Program.

Dzombak DA, Ali MA. 1993. Hydrochemical modeling of metal fate and transport in freshwater environments. *Water Pollut Res J Can* 28(1):7-50.

[EC] European Commission. 1996a. Technical guidance document in support of commission directive 93/67/EEC on risk assessment for new notified substances and commission regulation (EC) on risk assessment for existing substances, part II. Luxembourg: Office for Official Publications of the European Communities. nr1488/94.

[EC] European Commission. 1996b. EUSES documentation: The European union system for the evaluation of substances. NL: National Institute of Public Health and the Environment (RIVM). Available from the European Chemicals Bureau (EC/DGXI, ISPRA).

Endicott DD, Cook PM. 1994. Modeling the partitioning and bioaccumulation of TCDD and other hydrophobic organic chemicals in Lake Ontario. *Chemosphere* 28(1):75-87.

[EPRI] Electric Power Research Institute. 1996. EPRI's RIVRISK model: Assessing potential human health and ecological risks from chemical and thermal releases to rivers. Palo Alto CA, USA: EPRI. 5 p.

Erickson RJ, Benoit DA, Mattson VR. 1996. A prototype toxicity factors model for site-specific copper water quality criteria. Duluth MN, USA: USEPA, ERL.

Erickson RJ, Benoit DA, Mattson VR, Nelson Jr HP, Leonard EN. 1996. The effects of water chemistry on the toxicity of copper to fathead minnows. *Environ Toxicol Chem* 15(2):181-193.

Felmy AR. 1995. GMIN, a computerized chemical equilibrium program using a constrained minimization of the Gibbs free energy: Summary report. In: Loeppert R, Schwab AP, Goldberg S, editors. Chemical equilibrium and reaction models. Madison WI, USA: American Society of Agronomy. p 377-407.

Felmy AR, Brown SM, Onishi Y, Yabusaki SB, Argo RS. 1984. MEXAMS: The metals exposure analysis modeling system. Athens GA, USA: USEPA, ERL, ORD. NTIS PB84 157155.

Felmy AR, Girvin DC, Jenne EA. 1984. MINTEQ: A computer program for calculating aqueous geochemical equilibria. Athens GA, USA: USEPA, ERL, ORD. EPA-600-3-84-032.

Fisher NS, Reinfelder JR. 1995. The trophic transfer of metals in marine systems. In: *Metal Speciation and Bioavailability in Aquatic Systems*, Tessier A, Turner DR (editors). Chichester, UK: John Wiley and Sons. p 363-406.

Fisher NS, Stupakoff I, Sanudo-Wilhelmy SA, Wang W-X, Teyssie J-L, Fowler SW and Crusius J. 2000. Trace metals in marine copepods: A field test of a bioaccumulation model coupled to laboratory uptake kinetics data. *Mar Ecol Prog Series* 194:211-218.

Fisher NS, Wang W-X. 1998. Trophic transfer of silver to marine herbivores: A review of recent studies. *Environ Toxicol Chem* 17(4):562-571.

Forsythe BL, Klaine SJ, LaPoint TW, Cobb G, Bills T, Wenholz M. September 1997. Influence of water quality parameters on acute silver nitrate toxicity to 28-day old fathead minnows (*Pimephales promelas*). Final report Nr TIWET-9506.3. Submitted to the Silver Council c/o National Association of Photographic Manufacturers, by the Department of Environmental Toxicology, The Institute of Wildlife and Environmental Toxicology (TIWET), Clemson University, Pendleton, SC, USA.

Gailani J, Ziegler CK, Lick W. 1991. The transport of suspended solids in the Lower Fox River. *J Great Lakes Res* 17(4):479-494.

Garcia M, Parker G. 1991. Entrainment of bed sediment into suspension. *J Hydr Engrg* 117(4):414-435.

Garrels RM, Thompson ME. 1962. A chemical model for seawater at 25 C and one atmosphere total pressure. *Am J Sci* 260:57-66.

George SG. 1982. Subcellular accumulation and detoxication of metals in aquatic animals. In: Thurberg FP, Vernberg FJ, Vernberg WB, Calabrese A, editors. Physiological mechanisms of marine pollutant toxicity. Oxford, UK: Academic Press. p 3-52.

George SG. 1989. Cadmium effects on plaice liver xenobiotic and metal detoxication systems: Dose-response. *Aquat Toxicol* 15:303-310.

Gandhi N, Bhavsar S, Diamond ML. A coupled speciation and fate model applicable to mercury: BIOTRANSPEC. In preparation.

Gobas FAPC. 1993. A model for predicting the bioaccumulation of hydrophobic organic chemicals in aquatic food webs: Application to Lake Ontario. *Ecol Model* 69:1-17.

Gobas FAPC, Morrison HA. 2000. Bioconcentration and biomagnification in the aquatic environment. In: Boethling RS, MacKay D, editors. Handbook of property estimation methods for chemicals. Boca Raton FL, USA: Lewis Publishers, Environmental and Health Sciences. p 189-231.

Gobas FAPC, Z'graggen MN, Zhang X. 1995. Time response of the Lake Ontario ecosystem to virtual elimination of PCBs. *Environ Sci Technol* 29(8):2038-2046.

Gobas FAPC, Zhang X, Wells R. 1993. Gastrointestinal magnification: The mechanism of biomagnification and food chain accumulation of organic chemicals. *Environ Sci Technol* 27(13):2855-2863.

Grieb T, presenter. 1995. RIVRISK: A model to assess potential human health and ecological risks from power plant releases to rivers. Course notes presented by Tom Grieb at the Workshop on Aquatic Ecological Risk Assessment, Water Environment Federation Specialty Conference, Toxics Substances in Water Environments: Assessment and Control; 1995 May 14-17; Cincinnati, OH, USA.

Griscom SB, Fisher NS, Luoma SN. 2000. Geochemical influences on assimilation of sediment-bound metals in clams and mussels. *Environ Sci Tech* 34:91-99.

Griscom SB, Fisher NS. 2002. Uptake of dissolved Ag, Cd, and Co by the clam, *Macoma balthica*: Relative importance of overlying water, oxic pore water and burrow water. *Environ Sci Tech* 36:2471-2478.

Griscom SB, Fisher NS, Luoma SN. 2002. Kinetic modeling of Ag, Cd, and Co bioaccumulation in the clam *Macoma balthica*: Quantifying dietary and dissolved sources," *Mar Ecol Prog Series* 240:127-141.

Hamrick JM. 1992. A three-dimensional environmental fluid dynamics computer code: Theoretical and computational aspects. Gloucester Point VA, USA: The College of William and Mary, Virginia Institute of Marine Science. Special Report 317. 63 p.

Hamrick JM. 1996. User's manual for the environmental fluid dynamics computer code. In: Applied marine science and ocean engineering. Gloucester Point VA, USA: College of William and Mary, Virginia Institute of Marine Science, School of Marine Science, Department of Physical Sciences. Special Report nr 331.

Harrison SE, Klaverkamp JF. 1989. Uptake, elimination and tissue distribution of dietary and aqueous cadmium by rainbow trout (*Salmo gairdneri* Richardson) and lake whitefish (*Coregonus clupeaformis* Mitchell). *Environ Toxicol Chem* 8(1):87-97.

Heijerick DG, De Schamphelaere KAC, Janssen CR. 2002a. Predicting acute zinc toxicity for *Daphnia magna* as a function of key water chemistry characteristics: Development and validation of a biotic ligand model. *Environ Toxicol Chem* 21(6):1309-1315.

Heijerick DG, De Schamphelaere KAC, Janssen CR. 2002b. Biotic ligand model development predicting Zn toxicity to the algae *Raphidocelis subcapitata*: Possibilities and limitations. *Comp Biochem Physiol* Part C 133(1-2):207-318.

Hendriks AJ. 1995. Modelling non-equilibrium concentrations of microcontaminants in organisms: Comparative kinetics as a function of species size and octanol-water partitioning. *Chemosphere* 30(2):265-292.

Hildebrand CE, Tobey RA, Campbell EW and Enger MD. 1979. A cadmium-resistant variety of the Chinese hamster (CHO) cell with increased metallothionein induction capacity. *Exp Cell Res* 124:237-246.

Hook SE. May 2001. Sublethal toxicity of metals to copepods. A dissertation presented to the Graduate School in partial fulfillment of PhD in Coastal Oceanography, State University of New York (SUNY) at Stony Brook. 216 pp.

Hook SE, Fisher NS. 2002. Relating the reproductive toxicity of five ingested metals in calanoid copepods with sulfur affinity. *Mar Environ Res* 53:161-174.

Hudson RJM. 1998. Modeling the fate of metals in aquatic systems: The mechanistic basis of particle-water partitioning models. *Crit Rev Anal Chem* 28(2):19-26.

Hudson RJM, Gherini SA, Watras CJ, Porcella DB. 1994. A mechanistic model of the biogeochemical cycle of mercury in lakes. In: Watras CJ, Huckabee JW, editors. Mercury as a global pollutant. Chelsea MI, USA: Lewis.

HydroQual, Inc. 1981. CTAP documentation chemical transport analysis program. Mahwah, NJ, USA: HydroQual, Inc., HydroQual Report nr C109 for the Chemical Manufacturers Association Aquatic Research Task Group, Washington DC, USA. 110 p.

HydroQual, Inc. 1982a. Application guide for CMA-HydroQual chemical fate models. Mahwah, NJ, USA: HydroQual, Inc., HydroQual Report nr. C113 for the Chemical Manufacturers Association Aquatic Research Task Group, Washington DC, USA. 201 p.

HydroQual, Inc. 1982b. Application of an interactive water column-sediment model of heavy metals in the Clinton River. Warren MI, USA: General Motors Corporation. 144 p.

HydroQual, Inc. 1989. Validation of the general motors heavy metals water quality model of the Saginaw River. Warren MI, USA: General Motors Corporation. 31 p.

HydroQual, Inc. 1995. Development of total maximum daily loads and wasteload allocations for toxic metals in NY/NJ Harbor. Mahwah NJ, USA: HydroQual for USEPA, Region II, NY/NJ Harbor Estuary Program. 216 p.

HydroQual, Inc. 1997. Documentation for the probabilistic analysis program MONTE. Mahwah NJ, USA: HydroQual, Inc. 55 p.

HydroQual, Inc. 1998. A primer for ECOMSED, Version 1.3. Mahwah NJ, USA: HydroQual, Inc. 195 p.

HydroQual, Inc. 2003. Users guide for RCATOX. Mahwah NJ, USA: HydroQual, Inc.

Hydroscience, Inc. 1978. Estimation of PCB reduction by remedial action on the Hudson River ecosystem. Westwood NJ, USA: New York State Department of Environmental Conservation. 107 p.

Hydroscience, Inc. 1979. Analysis of the fate of PCB's in the ecosystem of the Hudson Estuary. Westwood NJ, USA: New York State Department of Environmental Conservation prepared by Hydroscience. 92 p.

Janes N, Playle RC. 1995. Modeling silver binding to gills of rainbow trout (*Oncorhynchus mykiss*). *Environ Toxicol Chem* 14(11):1847-1858.

Ji ZG, Hamrick JH, Pagenkopf J. 2002. Sediment and metals modeling in shallow river. *J Environ Engin* 128(2)105-119.

Jirka GH, Doneker RL, Hinton SW. 1996. User's manual for CORMIX: A hydrodynamic mixing zone model and decision support system for pollutant discharges into surface waters. Washington DC, USA: USEPA, Office of Science and Technology.

Johnson A, Luttik R. 1995. Risk assessment of antifoulants. Version 4. Presented at the 10th meeting of the Ad Hoc Group of Experts on Non-Agricultural Pesticides; 1995 Oct 2-3.

Johnson BH, Kim KW, Heath RE, Hsieh BB, Butler HL. 1993. Validation of three-dimensional hydrodynamic model of Chesapeake Bay. *J Hydrol Engrg* 119(1):2-20.

Jørgensen SE. 1990. A general model for the heavy metal pollution of aquatic ecosystems: Model development. *Environ Software* 5(3):136-141.

Kharaka YK, Barnes I. 1973. SOLMNEQ: Solution-mineral equilibrium computations. Springfield VA, USA: U.S. Department of Commerce. NTIS PB215-899.

Kieffer F. 1991. Metals and their compounds in the environment: Occurrence, analysis and biological relevance. Merian E, editor. New York NY, USA: Verlag Chemie. p 481-490.

Kinniburgh DG, Milne CJ, Benedetti MF, Pinheiro JP, Filius J, Koopal LK, Van Riemsdijk WH. 1996. Metal ion binding by humic acid: Application of the NICA-Donnan model. *Environ Sci Technol* 30(5):1687-1698.

Koopal LK, Van Riemsdijk WH, de Wit JCM, Benedetti MF. 1994. Analytical isotherm equations for multi-component adsorption to heterogeneous surfaces. *J Colloid Interface Sci* 166:51-60.

Kramer JR, Allen HE, Davison W, Godtfredsen KL, Meyer JS, Perdue EM, Tipping E, van de Meent D, Westall JC. 1997. Chemical speciation and metal toxicity in surface freshwaters. In: Bergman HL, Dorward-King EJ, editors. Reassessment of metals criteria for aquatic life protection: Priorities for research and implementation. Proceedings of the Pellston Workshop on Reassessment of Metals Criteria for Aquatic Life Protection; 1996 February 10-14; Pensacola FL, USA: SETAC.

Landrum PF, Reinhold MD, Nihart SR, Eadie BJ. 1985. Predicting the bioavailability of organic xenobiotics to *Pontoporeia hoyi* in the presence of humic and fulvic materials and natural dissolved organic matter. *Environ Toxicol Chem* 4:459-467.

Lee SS. 1979. A review of evaluative models in assessing the fate of pollutants. Menlo Park CA, USA: SRI International. USEPA. 38 p.

Lick W, Ziegler K, Tsai CH. 1987. Resuspension, deposition and transport of fine grained sediments in rivers and near shore areas. Santa Barbara CA, USA: University of California, Department of Mechanical and Environmental Engineering.

Lofts S, Tipping E. 1998. An assemblage model for cation binding by natural particulate matter. *Geochem Cosmochem Acta* 62(15):2609-2625.

LTI. 1992. Simplified Method Program-Variable Complexity Stream Toxics Model (SMPTOX3), Version 2.0, users manual. Athens GA, USA: USEPA, Region IV. (Updated in December 1992).

LTI. 1994. Dynamic toxics wasteload allocation model (DYNTOX). Version 2.0, users manual. Ann Arbor MI, USA: Limno-Tech.

Mackay D. 1991. Multimedia environmental models. Chelsea MI, USA: Lewis Publishers. 257 p.

Mackay D, Diamond ML. 1989. Application of the QWASI (Quantitative Water Air Sediment Interaction) fugacity model to the dynamics of organic and inorganic chemicals in lakes. *Chemosphere* 18:1343-1365.

Mackay D, Joy M, Paterson S. 1983. A quantitative water, air, sediment interaction (QWASI) fugacity model for describing the fate of chemicals in lakes. *Chemosphere* 12(9-10):981-997.

Mackay D, Paterson S. 1981. Calculating fugacity. *Environ Sci Technol* 15(9):1006-1014.

Mackay D, Paterson S. 1982. Fugacity revisited. *Environ Sci Technol* 16(12):654A-660A.

Mackay D, Paterson S, Joy M. 1983. A Quantitative Water, Air, Sediment Interaction (QWASI) fugacity model for describing the fate of chemicals in rivers. *Chemosphere* 12(9/10):1193-1208.

MacRae RK. 1994. The copper binding affinity of rainbow trout (*Oncorhynchus mykiss*) and brook trout (*Salvelinus fontinalis*) gills [MSc thesis]. Laramie WY, USA: Department of Zoology and Physiology and The Graduate School of the University of Wyoming.

MacRae RK, Smith DE, Swoboda-Colberg N, Meyer JS, Bergman HL. 1999. Copper binding affinity of rainbow trout (*Oncorhynchus mykiss*) and brook trout (*Salvelinus fontinalis*) gills. *Environ Toxicol Chem* 18:1180-1189.

Mancini JL. 1983. A method for calculating effects, on aquatic organisms, of time varying concentrations. *Water Res* 17(10):1355-1362.

Marr JCA, Hansen JA, Meyer JS, Cacela D, Podrabsky T, Lipton J, Bergman HL. 1998. Toxicity of cobalt and copper to rainbow trout: applications of a mechanistic model for predicting survival. *Aquat Toxicol* 43:225-238.

Martell AE, Motekaitis RJ. 1992. Determination and use of stability constants. New York NY, USA: VCH. 200 p.

Martin JL, Medine AJ. 1998. Metal exposure and transformation assessment model: model documentation for version 3. Boulder CO, USA: Medine Environmental Engineering, 110 pp.

Mason AZ, Jenkins KD. 1995. Metal detoxification in aquatic organisms. In: Tessier A, Turner DR, editors. Metal speciation and bioavailability in aquatic systems. Chichester, UK: John Wiley and Sons. p 479-608.

Masscheleyn PH, Patrick WH. 1993. Biogeochemical processes affecting selenium cycling in wetlands. *Environ Toxicol Chem* 12(12):2235-2243.

McCarty LS. 1987. Relationship between toxicity and bioconcentration for some organic chemicals. I. Examination of the relationship. In: Kaiser KLE, editor. QSAR in environmental toxicology, Volume II. Dordrecht NL: C. Reidel Publishing Company. p 207-220.

McCarty LS, Mackay D, Smith AD, Ozburn GW, Dixon DG. 1993. *Ecotoxicol Environ Saf* 25:253-270.

McGeer JC, Playle RC, Wood CM, Galvez F. 2000. A physiologically based biotic ligand model for predicting the acute toxicity of waterborne silver to rainbow trout in freshwaters. *Environ Sci Technol* 34(19):4199-4207.

Medine AJ. 1995. WASP4/META4. Cincinnati OH, USA: USEPA, Risk Reduction Engineering Laboratory. 72 pp.

Medine AJ, Bicknell BR. 1987. Case studies and model testing of the metals exposure analysis modeling system (MEXAMS). Athens GA, USA: USEPA, ERL, Research and Development. EPA-600-S3-86-045.

Medine AJ, Martin JL. 2000. Development of a metal exposure and transformation assessment model for use in watershed management, restoration and TMDL analyses. Proceedings of Watershed 2000, Water Environment Federation and British Columbia Water and Waste Association and Western Canada Water and Wastewater Association; 2000 July 8-12; Seattle WA. Washington DC, USA: Water Environment Foundation. 29 p.

Medine AJ, Martin JL, Sopher EJ. 2000. Development of the metal speciation-based metal exposure and transformation assessment model (META4): Application to copper and zinc problems in the Alamosa River, Colorado. In: RL Lipnick, RP Mason, ML Phillips, Pittman CV, editors. Chemicals in the environment. Fate, impacts, and remediation. Washington DC, USA: ACS. Symposium Series 80614.

Medine AJ, McCutcheon SC. 1987. Fate and transport of sediment-associated contaminants. In: Saxena J, editor. Hazard assessment of chemicals. Volume 6. New York NY, USA: Hemisphere Publishing. p 225-291.

Menzel DB. 1987. Physiological pharmacokinetic modeling. *Environ Sci Technol* 21(10):944-950.

Meyer JS, Gulley DD, Goodrich MS, Szmania DC, Brooks AS. 1995. Modeling toxicity due to intermittent exposure of rainbow trout and common shiners to monochloramine. *Environ Toxicol Chem* 14(1):165-175.

Meyer JS, Santore RC, Bobbitt JP, DeBrey LD, Boese CJ, Paquin PR, Allen HE, Bergman HL, Di Toro DM. 1999. Binding of nickel and copper to fish gills predicts toxicity when water hardness varies, but free-ion activity does not. *Environ Sci Technol* 33(6):913-916.

Mills WB, Dean JD, Porcella DB, Gherini SA, Hudson RJM, Frick WE, Rupp GL, Bowie GL. 1982a. Water quality assessment: A screening procedure for toxic and conventional pollutants. Parts 1, 2, and 3. Lafayette CA, USA: USEPA, prepared by Tetra Tech, Inc. EPA-600-6 82-004a, b, and c.

Mills WB, Dean JD, Porcella DB, Gherini SA, Hudson RJM, Frick WE, Rupp GL, Bowie GL. 1982b. Water quality assessment: A screening procedure for toxic and conventional pollutants. Part 2. Athens GA, USA: USEPA, ERL. EPA-600-6 82 004b.

Mills WB, Porcella DB, Ungs MJ, Gherini SA, Summers KV, Lingfung Mok, Rupp GL, Bowie GL, Haith DA. 1985. Water quality assessment: A screening procedure for toxic and conventional pollutants in surface and ground water. Part 1. Athens GA, USA: USEPA, ERL. EPA-600-6-85-002a.

Morel FM. 1983. Principles of aquatic chemistry. New York NY, USA: Wiley Interscience. 446 p.

Morel FM, Hering JG. 1993. Principles and applications of aquatic chemistry. New York NY, USA: John Wiley and Sons, Inc. 588 p.

Morel FM, Morgan JJ. 1972. Numerical method for computing equilibriums in aqueous chemical systems. *Environ Sci Technol* 6:58-67.

Mullenhoff WP, Soldate AM, Jr, Baumgartner DJ, Schuldt MD, Davis LR, Frick WE. 1985. Initial mixing characteristics of municipal ocean discharges: Volume I procedures and applications. Narragansett RI, USA: USEPA, ERL, ORD. NTIS Publication Number PB86-137478.

Nordstrom DK, Plummer LN, Wigley TML, Wolery TJ, Ball JW, Jenne EA, Bassett RL, Crerar DA, Florence TM, Fritz B, Hoffman M, Holdren GR, Lafon GM, Mattigod SV, McDuff RE, Morel F, Reddy MM, Sposito G, Thrailkill J. 1979. A comparison of computerized chemical models for equilibrium calculations in aqueous systems. In: Jenne EA, editor. Chemical modeling in aqueous systems. Washington DC, USA: ACS. p. 857-892.

Nott JA, Nicolaidou A. 1990. Transfer of metal detoxification along marine food chains. *J Mar Biol Assoc U K* 70:905-912.

O'Connor DJ. 1988a. Models of sorptive toxic substances in freshwater systems. I: Basic equations. *J Environ Engineering* 114(3). ASCE, ISSN Paper nr 22485.

O'Connor DJ. 1988b. Models of sorptive toxic substances in freshwater systems. II: Lakes and reservoirs. *J Environ Engineering* 114(3). ASCE, ISSN Paper nr 22486.

O'Connor DJ. 1988c. Models of sorptive toxic substances in freshwater systems. III: Streams and rivers. *J Environ Engineering* 114(3). ASCE, ISSN Paper nr 22487.

Ogden JM. 1984. Application of an interactive water column-sediment water quality model to the Saginaw River [thesis]. Available: GM Environmental Activities Staff and submitted to GMI Engineering and Management Institute, Michigan.

Oliver BG, Niimi A J. 1988. Trophodynamic analysis of polychlorinated biphenyl congeners and other chlorinated hydrocarbons in the Lake Ontario ecosystem. *Environ Sci Technol* 2(4):388-397.

Onishi Y, Thompson FL. 1984. Mathematical simulation of sediment and radionuclide transport in coastal seas. Volume 1: Testing of the sediment/radionuclide transport model, FETRA, NUREG/CR-2424; Richmond WA, USA: Batelle Pacific Northwest Laboratory. PNL-5088-1.

Onishi Y, Wise SE. 1982a. User's manual for the instream sediment-contaminant transport model SERATRA. Washington DC, USA: USEPA. EPA-600-3-82-055.

Onishi Y, Wise SE. 1982b. Mathematical model, SERATRA, for sediment-contaminant transport in Four Mile and Wolf Creeks in Iowa. Washington DC, USA: USEPA. EPA-600-3-82-045.

Pagenkopf GK. 1983. Gill surface interaction model for trace-metal toxicity to fishes: Role of complexation, pH, and water hardness. *Environ Sci Technol* 17(6):342 347.

Pagenkopf GK, Russo RC, Thurston RV. 1974. Effect of complexation on toxicity of copper to fishes. *J Fish Res Board Can* 31(4):462-465.

Paquin PR, Damiani D, Farley K, Santore RC, Di Toro DM. 2002. Development of a physiologically-based pharmacokinetic (PBPK) model of metal bioaccumulation by bivalves. Interactive poster presented at the 23rd Annual Meeting of the Society of Environmental Toxicology and Chemistry (SETAC), Salt Lake City, Utah, Abstract Book: IP50, p 149-150.

Paquin PR, Di Toro DM, Santore RC, Trivedi D, Wu KB. 1999. A biotic ligand model of the acute toxicity of metals. III. Application to fish and *Daphnia* exposure to silver. In: Integrated approach to assessing the bioavailability and toxicity of metals in surface waters and sediments. Washington DC, USA: USEPA, Science Advisory Board, OW, ORD. EPA-822-E-99-001. p 3-59 to 3-102.

Paquin PR, Gorsuch JW, Apte S, Batley GE, Bowles KC, Campbell PGC, Delos CG, Di Toro DM, Dwyer RL, Galvez F, Gensemer RW, Goss GG, Hogstrand C, Janssen CR, McGeer JC, Naddy RB, Playle RC, Santore RC, Schneider U, Stubblefield WA, Wood CM, Wu KB. 2002. The biotic ligand model: A historical overview. Special Issue: The Biotic Ligand Model for Metals–Current Research, Future Directions, Regulatory Implications, *Comp Biochem Physiol* Part C 133(1-2):3-35.

Paquin PR, Wu KB, Santore R, Di Toro DM. 1998. Application of a model of acute toxicity of silver to fish. Abstract Book for the 19th Annual Meeting of the Society of Environmental Toxicology and Chemistry (SETAC); 1998 Nov 15-18; Charlotte NC, USA.

Paquin PR, Zoltay V, Winfield RP, Wu KB, Mathew R, Santore RC, Di Toro DM. 2002. Extension of the biotic ligand model of acute toxicity to a physiologically-based model of the survival time of rainbow trout (*Oncorhynchus mykiss*) exposed to silver. Special Issue: The Biotic Ligand Model for Metals–Current Research, Future Directions, Regulatory Implications. *Comp Biochem Physiol* Part C 133(1-2):305-343.

Park RA. 1998. AQUATOX for windows: A modular toxic effects model for aquatic ecosystems, Eco Modeling, Montgomery Village MD. Washington DC, USA: USEPA, OW, OST. 64 p.

Parkhurst DL, Thorstenson DC, Plummer LN. 1980. PHREEQE-A computer program for geochemical applications. Reston VA, USA: USGS. WRI-80-96. 210 p.

Perkins EH, Kharaka YK, Gunter WD, DeBraal JD. 1990. Geochemical modeling of water-rock interactions using SOLMINEQ.88. In: Melchior DC, Bassett RL, editors. Chemical modeling of aqueous systems II. Washington DC, USA: ACS. p 117-127.

Phillips DJH. 1980. The effects of lipid on the accumulation of organochlorines and trace metals by biota. Chapter 3, Quantitative aquatic biological indicators: Their use to monitor trace metal and organochlorine pollution. London, UK: Applied Science Publishers Ltd. p 38-90.

Playle RC, Dixon DG, Burnison K. 1993a. Copper and cadmium binding to fish gills: Estimates of metal-gill stability constants and modelling of metal accumulation. *Can J Fish Aquat Sci* 50:2678-2687.

Playle RC, Dixon DG, Burnison K. 1993b. Copper and cadmium binding to fish gills: Modification by dissolved organic carbon and synthetic ligands. *Can J Fish Aquat Sci* 50:2667-2677.

Playle RC, Gensemer RW, Dixon DG. 1992. Copper accumulation on gills of fathead minnows: Influence of water hardness, complexation and pH on the gill micro-environment. *Environ Toxicol Chem* 11(3):381-391.

Plummer LN, Jones BF, Truesdell AH. 1976. WATEQF-A FORTRAN IV version of WATEQ, a computer program for calculation chemical equilibrium of natural waters. U.S. Geological Survey. Water Resource Investigations 76-13. 61 p.

Plummer LN, Parkhurst DL. 1990. Application of the Pitzer equations to the PHREEQE geochemical model. In: Melchior DC, Bassett RL, editors. Chemical modeling of aqueous systems II. Washington DC, USA: ACS. p 128-137.

Postek KM, Driscoll CT, Aber JD, Santore RC. 1995. Application of PnET-CN/CHESS to a spruce stand in Solling, Germany. *Ecol Model* 83(1-2):163-172.

Reinert KH, Rodgers JH. 1986. Validation trial of predictive fate models using an aquatic herbicide (Endothall). *Environ Toxicol Chem* 5:449-461.

Reinfelder JR, Fisher NS, Luoma SN, Wang W-X, 1998. "Trace element trophic transfer in aquatic organisms: A critique of the kinetic model approach," *Sci Tot Environ* 219:117-135.

Reinfelder JR, Wang WX, Luoma SN, Fisher NS. 1997. Assimilation efficiencies and turnover rates of trace elements in marine bivalves: A comparison of oysters, clams and mussels. *Mar Biol* 129:443-452.

Renner R. 1997. Rethinking water quality standards for metals toxicity. *Environ Sci Technol* 31(10):466A-468A.

[RIVM, VROM, WVC] National Institute of Public Health and Environmental Protection; Ministry of Housing, Special Planning and the Environment; Ministry of Welfare, Health and Cultural Affairs. 1994. Uniform system for the evaluation of substances (USES). Version 1.0. Amsterdam, NL: RIVM, VROM, WVC, The Hague, Ministry of Housing, Special Planning and the Environment. Distribution nr 11144/150.

Roesijadi G. 1992. Metallothioneins in metal regulation and toxicity in aquatic animals. *Aquat Toxicol* 22:81-114.

Roesijadi G, Klerks P. 1989. Kinetic analysis of Cd-binding to metallothionein and other intracellular ligands in oyster gills. *J Exp Zool* 251:1-12.

Roy RR, Campbell PGC. 1995. Survival time modeling of exposure of juvenile Atlantic salmon (*Salmo salar*) to mixtures of aluminum and zinc in soft water at low pH. *Aquat Toxicol* 33:155-176.

Ruiz CE, Schroeder PR, Aziz NM. 2000. Recovery: A contaminated sediment-water interaction model. Washington DC, USA: U.S. Army Corps of Engineers. ERDC/EL SR-D-00-1.

Salazar SM, Beckvar N, Salazar MH, Finkelstein K. 1996. An in-situ assessment of mercury contamination in the Sudbury River, Massachusetts, using bioaccumulation and growth in transplanted freshwater mussels. *NOAA Technical Memorandum NOS ORCA* 89:1-66.

Salazar MH, Salazar SM. 1995. In-situ bioassays using transplanted mussels: I. Estimating chemical exposure and bioeffects with bioaccumulation and growth. In: Hughes JS, Biddinger GR, Mones E, editors. Environ Toxicol Risk Assess - ASTM STP 1218. Philadelphia: American Society for Testing and Materials. 3:216-241.

Salazar MH, Salazar SM. 1998. Using caged bivalves as part of an exposure-dose-response triad to support an integrated risk assessment strategy. In: de Peyster A, Day K, editors. Chapter 8, Ecological risk assessment: A meeting of policy and science. Pensacola FL, USA. Society of Environmental Toxicology and Chemistry (SETAC): p 167-192.

Santore RS, Di Toro DM, Paquin PR, Allen HE, Meyer JS. 2001. A biotic ligand model of the acute toxicity of metals. II. Application to fish and *Daphnia* exposure to copper. *Environ Toxicol Chem* 20(10): 2397-2402.

Santore RC, Driscoll CT. 1995. The chess model for calculating chemical equilibria in soils and solutions. In: Loeppert R, Schwab AP, Goldberg S, editors. Chemical equilibrium and reaction models. Madison WI, USA: American Society of Agronomy. p 357-375.

Santore RC, Driscoll CT, Aloi M. 1995. A model of soil organic matter and its function in temperate forest soil development. In: McFee WW, Kelly JM, editors. Proceedings of the 8th North American Forest Soils Conference; 1993 May; Madison WI, USA: Soil Science Society of America. p 275-298.

Santore RC, McGrath J, Brix K, Paquin PR, Di Toro DM. 1998. Use of the biotic ligand model to calculate site specific water effect ratios for metals. Abstract Book for the 19th Annual Meeting of SETAC; 1998 Nov 15-19; Charlotte, NC, USA. Pensacola FL, USA: SETAC.

Santore RC, Mathew R, Paquin PR, Di Toro DM. 2002. Development of a biotic ligand model of acute toxicity for zinc. Special Issue: The Biotic Ligand Model for Metals–Current Research, Future Directions, Regulatory Implications. *Comp Biochem Physiol* Part C 133(1-2):271-285.

Schecher WD, McAvoy DC. 1992. MINEQL+: A software environment for chemical equilibrium modeling. *Comput Environ Urban Systems* 16:65-76.

Schlekat CE, Luoma SN. March-April 2000. You are what you eat: Incorporating dietary metals uptake into environmental quality guidelines for aquatic ecosystems. SETAC *Globe*, Learned Discourses: Timely Scientific Opinions 1(2):38-39.

Schnoor JL. 1996. Modeling trace metals. In: Environmental modeling: Fate and transport of pollutants in water, air and soil. New York NY, USA: John Wiley and Sons, Inc. p 381-454.

Schnoor JL, Connolly JP, Di Toro DM, de Rooij N, Diamond M, Kinerson RS, Porcella DB, Richardson WL, Stine JF. 1997. Environmental fate and transport. In: Bergman HL, Dorward-King EJ, editors. Reassessment of metals criteria for aquatic life protection: Priorities for research and implementation. Proceedings of the Pellston Workshop on Reassessment of Metals Criteria for Aquatic Life Protection; 1996 February 10-14; Pensacola FL, USA. Society of Environmental Toxicology and Chemistry (SETAC): 114 p.

Schnoor JL, Sato C, McKechnie D, Sahoo D. 1987. Processes, coefficients and models for simulating toxic organics and heavy metals in surface waters. Athens GA, USA: USEPA, ERL, ORD. EPA-600-3-87-015.

Serkiz S, Allison JD, Perdue EM, Allen HE, Brown DS. 1996. Correcting errors in the thermodynamic database for the equilibrium speciation model MINTEQA2. *Water Res* 30(8):1930-1933.

Shi B, Allen HE. 1995. Copper speciation and bioavailability: Critical evaluation for POTW effluent discharges. In: Toxics substances in water environments: Assessment and control. Proceedings of Water Environment Federation Specialty Conference; 1995 May 14-17; Cincinnati, OH, USA. Alexandria VA, USA: Water Environment Federation. 621 p.

Shi B, Allen HE, Grassi MT, Ma H. 1998. Modeling copper partitioning in surface waters. *Water Res* 32(12):3756-3764.

Sigg L. 1998. Partitioning of metals to suspended particles. In: Allen HE, Garrison AW, Luther III GW, editors. Metals in surface waters. Ann Arbor MI, USA: Ann Arbor Press. 262 p.

Smith RM, Martell AE, Motekaitis RJ. 1998. NIST critically selected stability constants of metal complexes database. Version 5.0. Gaithersburg MD, USA: U.S. Department of Commerce. 39 p. (NIST Standard Reference Database 46)

St. John J, Leo WM, Sheldon AW. 1985. Impact assessment of organotin chemicals in harbor environments. Ocean Engineering and the Environment, Marine Technology Society. Woodbridge NJ, USA: M&T Chemicals, prepared by HydroQual, Inc. 6 p.

Stephan CE, Mount DI, Hansen DJ, Gentile JH, Chapman GA, Brungs WA. 1985. Guidelines for deriving numerical national water quality criteria for the protection of aquatic organisms and their uses. Duluth MN, Narragansett RI, Corvallis OR, USA: USEPA, ORD, Environmental Research Laboratories.

Stumm W, Morgan JJ. 1981. Aquatic chemistry. New York NY, USA: John Wiley and Sons, Inc.

Suarez LA, Barber MC. 1994. FGETS version 3.0.18 user's manual. Athens GA, USA: USEPA.

Sunda WG, Guillard RRL. 1976. The relationship between cupric ion activity and the toxicity of copper to phytoplankton. *J Mar Res* 34(4):511-529.

Sunda WG, Hansen PJ. 1979. Chemical speciation of copper in river water: Effect of total copper, pH, carbonate, and dissolved organic matter. In: Jenne EA, editor. Chemical modeling in aqueous systems. Washington DC, USA: ACS. Symposium Series 93. p 147-180.

Suter GW, editor. 1993. Ecological risk assessment. Ann Arbor MI, USA: Lewis Publishers.

Szumski DS, Barton DA. 1983. Development of a mechanistic model of acute heavy metal toxicity. Aquatic Toxicology and Hazard Assessment: 6th Symposium. Philadelphia PA, USA: ASTM. Special Technical Publication 802. Publication Code Number (PCN) 04-802000-16. p 42-72.

Temminghoff EJM, Van der Zee SEATM, de Haan FAM. 1997. Copper mobility in a copper-contaminated sandy soil as affected by pH and solid and dissolved organic matter. *Environ Sci Technol* 31(4):1109-1115.

Tetra Tech, Inc. 2002a. Theoretical and computational aspects of sediment and contaminant transport in the EFDC Model. Fairfax VA, USA: Tetra Tech, Inc. 3rd Draft of an EFDC Technical Memorandum prepared for USEPA OST, Washington DC, USA:

Tetra Tech, Inc. 2002b. Draft user's manual for Environmental Fluid Dynamics Code, Hydro Version (EFDC-Hydro). Release 1.00. FairfaxVA, USA: Prepared for USEPA Region 4, Atlanta GA, by Tetra Tech, Inc.

Thomann RV. 1977. A trophic length model of the fate of hazardous substances in the aquatic food chain. Westwood NJ, USA: Hydroscience, Inc. 53 p.

Thomann RV. 1978. Size dependent model of hazardous substances in aquatic food chains. Duluth, URD. Duluth MN, USA: USEPA, ERL. Ecological Research Series. EPA-600-3-78-036. 40 p.

Thomann RV. 1981. Equilibrium model of the fate of microcontaminants in diverse aquatic food chains. *Canad J Fish Aquat Sci* 38:280-296.

Thomann RV, Connolly JP. 1984. Model of PCB in the Lake Michigan trout food chain. *Environ Sci Technol* 18(2):65-71.

Thomann RV, Connolly JP, Parkerton TF. 1992a. Modeling accumulation of organic chemicals in aquatic food webs. In: Gobas FAPC, McCorquodale JA, editors. Chemical dynamics in fresh water ecosystems. Boca Raton FL, USA: Lewis Publishers. 247 p.

Thomann RV, Connolly JP, Parkerton TF. 1992b. An equilibrium model of organic chemical accumulation in aquatic food webs with sediment interactions. *Environ Toxicol Chem* 11(5):615-629.

Thomann RV, Di Toro DM. 1983. Physico-chemical model of toxic substances in the Great Lakes. *J Great Lakes Res* 9(4):474-496.

Thomann RV, Mahony JD, Mueller R. 1995. Steady-state model of biota sediment accumulation factor for metals in two marine bivalves. *Environ Toxicol Chem* 14(11):1989-1998.

Thomann RV, Mueller JA. 1987. Principles of surface water quality modeling and control. New York NY, USA: Harper and Rowe Publishers. 644 p.

Thomann RV, Mueller JA, Winfield RP, Huang CR. 1991. Model of the fate and accumulation of PCB homologues in Hudson Estuary. *ASCE J Environ Engr* 117(2):161-177.

Thomann RV, Shkreli F, Harrison S. 1997. A pharmacokinetic model of cadmium in rainbow trout. *Environ Toxicol Chem* 16(11):2268-2274.

Thomann RV, Snyder CA, Squibb KS. 1994. Development of a pharmacokinetic model for chromium in the rat following subchronic exposure: I. The importance of incorporating a long-term storage compartment. *Toxicol App Pharm* 128:189-198.

Thomann RV, Szumski DS, Di Toro DM, O'Connor DJ. 1974. A food chain model of cadmium in Western Lake Erie. *Water Research* 8:841-849.

Tipping E. 1993. Modeling the competition between alkaline earth cations and trace metal species for binding by humic substances. *Environ Sci Technol* 27(3):520-529.

Tipping E. 1994. WHAM-a chemical equilibrium model and computer code for waters, sediments and soils incorporating a discrete site/electrostatic model of ion binding by humic substances. *Comput Geosci* 20(6):973-1023.

Tipping, E. 1997. Humic ion-binding model VI: An improved description of the interactions of protons and metal ions with humic substances. *Aquat Geochem* 4:3-48.

Tipping E, Backes CA, Hurley MA. 1988. The complexation of protons, aluminum and calcium by aquatic humic substances: A model incorporating binding-site heterogeneity and macroionic effects. *Water Res* 22(5):597-611.

Tipping E, Backes CA, Hurley MA. 1989. Modeling the interactions of Al species, protons, and Ca with humic substances in acid waters and soils. In: Lewis TE, editor. Environmental chemistry and toxicology of aluminum. London: Lewis Publishers. p 59-82.

Tipping E, Hurley MA. 1988. A model of solid-solution interactions in acid organic soils, based on the complexation properties of humic substances. *J Soil Sci* 39:505-519.

Tipping E, Reddy MM, Hurley MA. 1990. Modeling electrostatic and heterogeneity effects on proton dissociation from humic substances. *Environ Sci Technol* 24(11):1700-1705.

Tipping E, Woof C. 1990. Humic substances in acid organic soils: Modelling their release to the soil solution in terms of humic charge. *J Soil Sci* 41:573-586.

Tipping E, Woof C. 1991. The distribution of humic substances between the solid and aqueous phases of acid organic soils: A description based on humic heterogeneity and charge-dependent sorption equilibria. *J Soil Sci* 42:437-448.

Tipping E, Woof C, Hurley MA. 1991. Humic substances in acid surface waters: Modelling aluminium binding, contribution to ionic charge-balance, and control of pH. *Water Res* 25(4):425-435.

Tischler L. 1998. Waste load allocations for metals. In: Allen HE, Garrison AW, Luther III GW, editors. Metals in surface waters. Ann Arbor MI, USA: Ann Arbor Press. p 67-89.

Truesdell AH, Jones BJ. 1973. WATEQ, a computer program for calculating chemical equilibria of natural waters. Springfield VA, USA: NTIS. PB2-20464

Tsai CH, Lick W. 1987. Resuspension of sediments from Long Island Sound. *Water Sci Technol* 21(6/7):314-321.

[USEPA] U.S. Environmental Protection Agency. 1987. Selection criteria for mathematical models used in exposure assessments: Surface water models. Washington DC, USA: USEPA, Exposure Assessment Group, Office of Health and Environmental Assessment. EPA-600-8-87-042.

[USEPA] U.S. Environmental Protection Agency. 1989. Briefing report to the EPA Science Advisory Board on the equilibrium partitioning approach to generating sediment quality criteria. Washington DC, USA: USEPA, OW Regulations and Standards, Criteria and Standards Division. EPA-440-5-890-002.

[USEPA] U.S. Environmental Protection Agency. 1991. MINTEQA2 user manual. Version 3.11. Athens GA, USA: USEPA, Center for Exposure Assessment Modeling, ERL. 106 p.

[USEPA] U.S. Environmental Protection Agency. 1993. Technical basis for deriving sediment quality criteria for nonionic organic contaminants for the protection of benthic organisms by using equilibrium partitioning. Washington DC, USA: USEPA, OW. EPA-822-R-93-011.

[USEPA] U.S. Environmental Protection Agency. 1994a. Water quality standards handbook. 2nd ed. Appendix L, Interim guidance on determination and use of water effect ratios for metals. Washington DC, USA: USEPA, OW. EPA-823-B-94-005a.

[USEPA] U.S. Environmental Protection Agency. 1994b. Briefing report to the EPA Science Advisory Board on the equilibrium partitioning approach to predicting metal bioavailability in sediments and the derivation of sediment quality criteria for metals. Washington DC, USA: USEPA, OW and ORD. EPA-822-D-94-002.

[USEPA] U.S. Environmental Protection Agency. 1997. Compendium of tools for watershed assessment and TMDL development. Washington DC, USA: USEPA, Watershed Branch, Assessment and Watershed Division, Office of Wetlands, Oceans and Watersheds. EPA-841-B-97-006.

[USEPA] U.S. Environmental Protection Agency. 1998a. Ambient water quality criteria derivation methodology human health: technical support document. Washington DC, USA: USEPA. EPA-822-B-98-005. Final draft.

[USEPA] U.S. Environmental Protection Agency. 1998b. An SAB report: Evaluation of the Blackstone River initiative. Washington DC, USA: USEPA, prepared by the Ecological Processes and Effects Committee. EPA-SAB-EPEC-98-011.

[USEPA] U.S. Environmental Protection Agency. 1999a. Integrated approach to assessing the bioavailability and toxicity of metals in surface waters and sediments. Washington DC, USA: USEPA, Science Advisory Board, OW, ORD. EPA-822-E-99-001.

[USEPA] U.S. Environmental Protection Agency. 1999b. Partition coefficients for metals in surface water, soil, and waste. Athens GA, USA: USEPA, ORD, prepared by HydroGeoLogic, Inc. and Allison Geoscience Consultants, Inc. Available at www.epa.gov/epaoswer/hazwaste/id/hwirwste/pdf/risk/reports/s0524.pdf. 51 p. Accessed 30 May 2003.

[USEPA] U.S. Environmental Protection Agency. 1999c. Understanding variation in partition coefficient, K_d, values. Volumes I and II. Washington DC, USA: USEPA, Office of Radiation and Indoor Air and Office of Environmental Restoration. EPA-402-R-99-004A. Available at www.epa.gov/rpdweb00/cleanup/partition. Accessed 23 Sep 2003.

[USEPA] U.S. Environmental Protection Agency. 2000. An SAB report: Review of the biotic ligand model of the acute toxicity of metals. Washington DC, USA: USEPA, Ecological Processes and Effects Committee of the Science Advisory Board. EPA-SAB-EPEC-00-0006.

[USEPA] U.S. Environmental Protection Agency, Electric Power Research Institute (EPRI), and the Utility Water Act group. 1996. The metals translator: Guidance for calculating a total recoverable permit limit from a dissolved criterion. Washington DC, USA: USEPA, OW. EPA-823-B-96-007.

Verhaar HJM, de Wolfe W, Dyer S, Legierse KCHM, Seinen W, Hermens JLM. 1999. An LC50 vs time model for the aquatic toxicity of reactive and receptor-mediated compounds. Consequences for bioconcentration kinetics and risk assessment. *Environ Sci Technol* 33(5):758-763.

Wallace WG, Lee B-G, Luoma SN. 2003. Subcellular compartmentalization of Cd and Zn in two bivalves. I. Significance of metal-sensitive fractions (MSF) and biologically detoxified metal (BDM). *Mar Ecol Prog Series* 249:183-197.

Wallace WG, Lopez GR. 1996. Relationship between subcellular cadmium distribution in prey and cadmium trophic transfer to a predator. *Estuaries* 19(4):923-930.

Wallace WG, Lopez GR. 1997. Bioavailability of biologically sequestered cadmium and the implications of metal detoxification. *Mar Ecol Prog Ser* 147:149-157.

Wallace WG, Lopez GR, Levinton JS. 1998. Cadmium resistance in an oligochaete and its effect on cadmium trophic transfer to an omnivorous shrimp. *Mar Ecol Prog Series* 172:225-237.

Wallace WG, Luoma SN. 2003. Subcellular compartmentalization of Cd and Zn in two bivalves. II. Significance of trophically available metal (TAM). *Mar Ecol Prog Series* (forthcoming).

Wang W-X, Fisher NS, Luoma SN. 1995. Assimilation of trace elements ingested by the mussel *Mytilus edulis*: Effects of algal food abundance." *Mar Ecol Prog Series* 129:165-176.

Wang W-X, Reinfelder JR, Lee BG, Fisher NS. 1996. Assimilation and regeneration of trace elements by marine copepods. *Limnol Ocean* 41(1):70-81.

Wang, W-X, Fisher NS. 1996a. Assimilation of trace elements by the mussel *Mytilus edulis*: effects of diatom chemical composition. *Mar Bio* 125:715-724.

Wang W-X, Fisher NS. 1996b. Assimilation of trace elements and carbon by the mussel *Mytilus edulis*: effects of food composition. *Limnol Oceanog* 41(2):197-207.

Wang W-X, Griscom SB, Fisher NS. 1997. Bioavailability of Cr(III) and Cr(VI) to Marine mussels from solute and particulate pathways. *Environ Sci Tech* 31(2):603-611.

Wang W-X, Ke C. 2002. Dominance of dietary intake of cadmium and zinc by two marine predatory gastropods. *Aquatic Toxicology* 56:153-165.

Waybrant R. 1973. Factors controlling the distribution and persistence of lindane and DDE in lentic environments. Ann Arbor MI, USA: University Microfilms, Purdue University. Order No. 74-15, 256. 175 p.

Westall J. 1979. MICROQL: I. A chemical equilibrium program in BASIC. Duebendorf, Switzerland: Swiss Federal Institute of Technology EAWAG CH-8600.

Westall J, Hohl H. 1980. A comparison of electrostatic models for the oxide/solution interface. *Adv Colloid Interface Sci* 12:265-294.

Westall JC, Zachary JL, Morel FMM. 1976. MINEQL: A computer program for the calculation of chemical equilibrium composition of aqueous systems. Cambridge MA, USA: Massachusetts Institute of Technology, Department of Civil Engineering, Ralph M Parsons Laboratory. 91 p.

Winge D, Krasno J, Colucci AV. 1974. Cadmium accumulation in rat liver: correlation between bound metal and pathology. In Hoekstra WG, Suttie JW, Ganther HE, Mertz W, editors. Trace element metabolism in animals – 2. Baltimore MD, USA: University Park Press. p 500-502.

Wolery TJ. 1979. Calculation of chemical equilibrium between aqueous solution and minerals: The EQ3/6 software package. Livermore CA, USA: Lawrence Livermore National Laboratory. UCRL-52658.

Wolery TJ, Jackson KJ, Bourcier WL, Bruton CJ, Viani BE, Knauss KG, Delany JM. 1990. Current status of the EQ3/6 software package for geochemical modeling. In: Melchior DC, Bassett RL, editors. Chemical modeling of aqueous systems II. Washington DC, USA: ACS. p 104-116.

Wright RM. 1987. Development of a one-dimensional water quality model for the Blackstone river, part 2: Mathematical modeling. Providence RI, USA: RIDEM.

Wright RM, Nolan PM, Pincumbe D, Hartman E. 1998. Blackstone River initiative: Water quality analysis of the Blackstone River under wet and dry weather conditions. Boston MA, USA: U.S. Environmental Protection Agency, New England. 527 p.

Yanai RD, Santore RC. 1996. YASE soil equilibrium model. Palo Alto CA, USA: EPRI.

Yanai RD, Zollweg CG, Santore RC. 1996. Using the soil model YASE alone or in combination with TREGRO. Model description and user's guide, Version 3.0.2. Palo Alto CA, USA: EPRI.

Yeh GT. 1981. ICM, An integrated compartment method for numerically solving partial differential equations. Oak Ridge TN, USA: Oak Ridge National Laboratory. ORNL-5684.

Yeh GT. 1982. CHNTRN, A channel transport model for simulating sediment and chemical distribution in a stream/river network. Oak Ridge TN, USA: Oak Ridge National Laboratory. ORNL-5882.

Ziegler CK, Lick W. 1986. A numerical model of the resuspension, deposition and transport of fine-grained sediments in shallow water. Santa Barbara CA, USA: University of California Santa Barbara. Report ME-86-3.

Zillioux EJ, Porcella DB, Benoit JM. 1993. Mercury cycling and effects in freshwater wetland ecosystems. *Environ Toxicol Chem* 12(12):2245-2264.

Index

A

Abbreviations, 127–132
Acid-volatile sulfide (AVS), 10, 22, 25, 62
Acute toxicity model, 100
Advection–dispersion equation, 27
Alosa pseudoharengus, 69
Amphipods, 83, 84
Analytical solution models, 15, 16, 18
Antifoulant(s)
    leaching rate of, 22
    risk assessment for, 22
Aquatic fate and transport models, 5–11, 13–52
    literature review, 11
    literature search, 14
    model frameworks applied to exposure and risk assessments, 6–10
    model reviews and guidance documents, 51–52
    review of models, 14–50
        comparison of model features, 35–50
        overview and description of models, 14–35
Aquatic food web models, 62
AQUATOX, 68, 106
Atmosphere submodel, 22
AVS, *see* Acid-volatile sulfide

B

BASS model, 68
BCF, *see* Bioconcentration factor
Bioaccumulation
    equation for, 63
    modeling, difficulty in, 63
    potential, 1–2
Bioaccumulation and toxicity models, 61–90
    bioaccumulation models, 62–73
        application of bioaccumulation models, 67–73
        food web, 70
        general equation for bioaccumulation, 63–65
        sources of, 106
    integration of bioaccumulation and toxicity models for metals, 87–90
    modeling metal toxicity in sediments, 81–86
    toxicity models for waterborne metals, 73–81
Bioconcentration factor (BCF), 61
Biota–sediment
    accumulation factor (BSAF), 61
    ratios, 70
Biotic Ligand Model (BLM), 2, 9, 50, 106
    ability of to predict copper toxicity, 76
    conceptual diagram of, 76
    development of, 97, 100
    framework
        important feature of, 75
        testing of, 10
    metal bioavailability determined using, 54
    metal–organic matter interaction and, 94
    metal speciation computations, 9
    metal toxicity in, 76
    technical basis for, 78
    use of in risk assessments, 62
BIOTRANSPEC model, 21
Bioturbation, particle mixing due to, 44
Bivalves, pathways for metal uptake by, 72
Black boxes, treatment of models as, 58
BLM, *see* Biotic Ligand Model
BSAF, *see* Biota–sediment accumulation factor

C

Cadmium, pharmacokinetic model for, 74
CAUC, *see* Critical area under the curve
CBR model, *see* Constant Body Residue model
CCOD, *see* Current Content on Diskette
Channel Transport (CHNTRN) model, 31
Chemical(s)
    binding, electrostatic interactions and, 56
    concentration–response curves for nonionic organic, 82
    hydrophobic organic, 62, 95
    partitioning, 44, 49
    precipitation, 32
    speciation
        compact numerical algorithm for, 55
        database, 17
    transfers, 47
    water column–bed transfer rates of, 48
Chemical equilibrium models, 2, 47, 53–60
    comparison of models, 57–60
    features of, 58
    historical model development, 53–57
    sources of, 105

Chemical Equilibrium in Soils and Solutions (CHESS), 56, 75, 94, 105
Chemical Manufacturers Association (CMA), 19
Chemical Transport and Analysis Program (CTAP), 23, 93, 103
CHESS, see Chemical Equilibrium in Soils and Solutions
CHNTRN model, see Channel Transport model
Chromium, tissue accumulation of, 74
Clear Creek Superfund Site, 33
CMA, see Chemical Manufacturers Association
CMV, see Completely mixed volume
Cobalt–copper mixtures, overprediction of toxicity in, 80
Completely mixed volume (CMV), 39
Computer codes, equilibrium models evolved from, 54
Constant Body Residue (CBR) model, 80
Contaminant
 passive diffusion of through GIT membrane, 65
 transfer, 63
Copper
 BLM-predicted LC50 versus observed LC50 for, 77
 –DOC interactions, 33
 toxicity, 76, 100
 WQC for, 47, 96
CORMIX, 39
Crassostrea virginica, 71, 72
Critical area under the curve (CAUC), 80
Critical body residue, 10, 96, 100
CTAP, see Chemical Transport and Analysis Program
Current Content on Diskette (CCOD), 14
Cytochrome P-450, 101

D

Damage-repair model, 80
Danish Hydraulic Institute (DHI), 34
Daphnia
 magna, 78, 96
 pulex, 78, 96
Database
 chemical speciation, 17
 thermodynamic, 94
 WHAM, 94
Daughter product, 37
DDE, see Dichlorodiphenyldichloroethylene

DELFT3D, 30, 103
Depuration kinetics, 71, 100
DHI, see Danish Hydraulic Institute
DIALOG, 14
Dichlorodiphenyldichloroethylene (DDE), 19, 48
Dioxin-containing discharge, 19
Dissolved organic carbon (DOC), 7, 26, 69, 99
Dissolved organic matter (DOM), 7, 54, 57
DJOC, 103
DOC, see Dissolved organic carbon
DOM, see Dissolved organic matter
Donnan-type diffuse layer model, 56
Dynamic Toxics (DYNTOX) model, 34
Dynamic water quality models, 52
DYNTOX model, see Dynamic Toxics model

E

EC, see European Commission
ECOM, see Estuary, Coastal, Ocean Model
ECOMSED, see Estuary, Coastal, Ocean Model-Sediment
Electric Power Research Institute (EPRI), 17
Electric power utilities, 19
Electrostatic adsorption models, 58
Electrostatic interactions, chemical binding and, 56
EPRI, see Electric Power Research Institute
Equilibrium
 assumptions, 41
 constants, values for, 57
 partitioning-based sediment guideline (ESG), 81, 82
Equivalent mortality dose, 78
EROD, see Ethoxyresorufin-O-dealkylase
ESG, see Equilibrium partitioning-based sediment guideline
Estuary, Coastal, Ocean Model (ECOM), 30, 92
Estuary, Coastal, Ocean Model-Sediment (ECOMSED), 30, 92
Ethoxyresorufin-O-dealkylase (EROD), 101
European Commission (EC), 1
European Commission technical guidance document (TGD), 1, 22
 partition coefficient, 45
 recommendation, 50
European Union System for the Evaluation of Substances (EUSES), 22, 104
EUSES, see European Union System for the Evaluation of Substances

Eutrophication, 27
EXAMS, *see* Exposure Analysis Modeling System
EXAMS II, 103
Exposure Analysis Modeling System (EXAMS), 9, 19, 25, 26, 27

F

Fate and transport model(s)
    chemical equilibrium computation, 48
    chemical transfers in, 47
    features of, 36–37
    limitation of, 93
    sources of, 103–105
FGETS model, *see* Food and Gill Exchange of Toxic Substances model
FIAM, *see* Free-ion activity model
Field survey conditions, 11
First-order decay, 9, 19, 31, 37, 50
Floating point precision, 58
Fluid transport, 14
    analysis, 6
    regimes, 35, 38, 39
Food
    contaminated, 65
    ingestion, 62, 65
    metal assimilation efficiency from, 71
Food chain
    Great Lakes, 67, 95
    model, multicompartment, 65, 66
    submodels, 22
Food and Gill Exchange of Toxic Substances (FGETS) model, 67, 68
Food web bioaccumulation model, 70
Free energy minimization, 54
Free-ion activity model (FIAM), 75

G

Gastrointestinal tract (GIT), 65
Geochemical model, 51
Geochemical submodel, 97
Gill surface interaction model, 96
GIT, *see* Gastrointestinal tract
Gobas model, 67, 68, 96
Graphical user interface, 55
Great Lakes food chain, PCBs in, 67, 95
Groundwater, mineral equilibria in, 55

H

HOCs, *see* Hydrophobic organic chemicals
HSPF, *see* Hydrologic Simulation Program-FORTRAN

Hydrodynamic models, 52, 92
Hydrologic Simulation Program-FORTRAN (HSPF), 30, 31, 103
Hydrolysis, 37
Hydrophobic organic chemicals (HOCs), 62, 95
    bioaccumulation of, 73
    biomagnification of, 65
    coexistence of metals and, 101
    concentrations, biomagnified, 67
HydroQual, 14

I

IBM, *see* Ion Balance Model
ILL, *see* Incipient lethal level
Incipient lethal level (ILL), 80
Intracellular speciation measurements, 90
Ion activity coefficients, 58
Ion Balance Model (IBM), 81, 97

L

Level of no concern, application of statistical methods to define, 2
Lindane, 19, 48
Lipid normalization, 89

M

MCM, *see* Mercury Cycling Model
Mercury Cycling Model (MCM), 9
META4, *see* Metal Exposure and Transformation Assessment Model
Metal(s)
    acute toxicity model for, 100
    bioavailability
        determination of using BLM, 54
        evaluation of, 3
    bioreactive, 89
    calibration of WHAM to, 60
    disappearance rate of, 22
    distribution of, 6, 73
    DOM interactions, WHAM and, 94, 99
    models representing characteristics of, 3
    –organic matter interactions, calibration of, 94
    organism exposure to, 71
    risk assessment, 98
    speciation
        characterization of, 2
        chemical equilibrium model for, 53, 59
        computations, 9
        stand-alone model to evaluate, 51

Metal(s) *cont*
　　toxicity, spillover hypothesis of, 89
　　transformation processes, algorithms for simulation of, 32
　　trophic transfer potential of accumulated, 87
　　waterborne exposure to, 88
Metal Exposure and Transformation Assessment Model (META4), 32
Metallothionein (MT), 70, 89
Metals Exposure Analysis Modeling Systems (MEXAMS), 25
MEXAMS, *see* Metals Exposure Analysis Modeling Systems
MINEQL+, 105
MINTEQ, 26
MINTEQA2, 105
Mixing zone models, 52
Model(s)
　　acute toxicity, 100
　　analytical solution, 15, 16, 18
　　aquatic food web, 62
　　AQUATOX, 68, 106
　　assumptions, 11
　　availability of, 13, 35, 50
　　BASS, 68
　　bioaccumulation, *see* Bioaccumulation and toxicity models
　　Biotic Ligand (BLM), 2, 9, 50, 106
　　　　ability of to predict copper toxicity, 76
　　　　conceptual diagram of, 76
　　　　development of, 97, 100
　　　　framework, 10, 75
　　　　metal bioavailability determined using, 54
　　　　metal–organic matter interaction and, 94
　　　　metal toxicity in, 76
　　　　technical basis for, 78
　　　　use of in risk assessments, 62
　　BIOTRANSPEC, 21
　　calibration, 11
　　Channel Transport (CHNTRN), 31
　　chemical equilibrium, 2, 47, 53–60
　　　　comparison of models, 57–60
　　　　features of, 58
　　　　historical model development, 53–57
　　　　sources of, 105

Chemical Equilibrium in Soils and Solutions (CHESS), 56, 75, 94, 105
Chemical Transport and Analysis Program (CTAP), 23, 93, 103
Constant Body Residue (CBR), 80
CORMIX, 39
damage-repair, 80
DELFT3D, 30, 103
dimensionality, 40
discrepancies between, 57
DJOC, 103
Donnan-type diffuse layer, 56
Dynamic Toxics (DYNTOX), 34
dynamic water quality, 52
electrostatic adsorption, 58
Estuary, Coastal, Ocean Model (ECOM), 30, 92
Estuary, Coastal, Ocean Model-Sediment (ECOMSED), 30, 92
European Union System for the Evaluation of Substances (EUSES), 22, 104
EXAMS II, 103
Exposure Analysis Modeling System (EXAMS), 9, 19, 25, 26, 27
Food and Gill Exchange of Toxic Substances (FGETS), 67, 68
free-ion activity, 75
geochemical, 51
gill surface interaction, 96
Gobas, 67, 68, 96
hydrodynamic, 52, 92
Hydrologic Simulation Program-FORTRAN (HSPF), 30, 31, 103
Ion Balance (IBM), 81, 97
Mercury Cycling (MCM), 9
Metal Exposure and Transformation Assessment (META4), 32
MINEQL+, 105
MINTEQA2, 105
mixing zone, 52
multicompartment food chain, 65, 66
Nonideal Competitive Adsorption (NICA), 47, 57, 95
nutrient cycling, 56
physiologically based pharmacokinetic, 73, 81
PLUMES, 39
Probabilistic Dilution (PDM), 34
Quantitative Water Air Sediment Interaction (QWASI), 20, 21, 35, 103

RECOVERY, 104
RIVEQLII, 42
River Risk (RIVRISK), 17, 35, 42, 104
Row-Column AESOP for Toxics (RCATOX), 29
sediment chemistry, 99
sediment diagenesis, 30
sediment transport, 29, 91, 92
Simplified Lake and Stream Analysis (SLSA), 19, 20, 93, 104
Simplified Method Program-Variable-Complexity Stream Toxics (SMPTOX), 25, 104
simultaneously extracted metal–acid-volatile sulfide (SEM–AVS), 99
stand-alone, 11, 51, 39
Stanford Watershed, 31
steady-state, 16, 23, 24, 52
Thomann, 67, 68, 96, 106
time variable numerical solution, 16, 26, 28
Uniform System for the Evaluation of Substances (USES), 22, 35, 104
validation tests, 23
Water Quality Analysis Simulation Program (WASP), 27
Water Quality Analysis Simulation Program Version 5 (WASP5), 9, 104
Water Quality Analysis Simulation of Toxics (WASTOX), 29, 44, 93, 104
Water Quality Assessment Methodology (WQAM), 17, 51, 93, 105
Windermere Humic Aqueous Model (WHAM), 47, 55, 75, 94, 105
    calibration of to variety of metals, 60
    lack of generic computation framework in, 56
    metal–DOM interactions and, 99
    metal–organic matter interactions in, 56
Modeling assumptions, uncertainty in, 1
Model selection and future model development needs, 91–102
    future model development needs, 97–101
    bioaccumulation model, 100–101
    chemical equilibrium model, 98–99
    dynamic simulations and probabilistic overlay, 98

    sediment chemistry model, 99
    toxicity model, 100
model selection, 91–97
    bioaccumulation and toxicity models, 95–97
    chemical equilibrium models, 94–95
    fate and transport models, 93
    hydrodynamic models, 92
    sediment transport models, 92
Monte Carlo analysis, 34
Monte Carlo generation of inputs, 98
Monte Carlo simulation techniques, 23
MT, *see* Metallothionein
Mytilus edulis, 71, 72, 96

N

National Oceanic and Atmospheric Administration (NOAA), 71
Near field mixing zone, 41
NICA model, *see* Nonideal Competitive Adsorption model
NOAA, *see* National Oceanic and Atmospheric Administration
NOEC, *see* No-observed-effect concentration
Nonideal Competitive Adsorption (NICA) model, 47, 57, 95
No-observed-effect concentration (NOEC), 1
Nutrient cycling models, 56

O

Oncorhynchus kisutch, 69
Organic contaminants, accumulation of in organisms, 69
Organophosphorus, 89
Oxidation, 37

P

PAHs, *see* Polyaromatic hydrocarbons
Particle
    settling, 25, 44
    transfers, 48
Particulate
    adsorption, 58
    organic carbon (POC), 45, 99
    transport, 6, 24, 28
        mechanisms, 42
        representations, comparison of alternative, 43
Partition coefficient, 52

Partitioning
    approaches, 45, 46
    evaluation of, 50
    importance of, 7–9
    reactions, 7
PAWTOXIC, see Pawtuxent Toxics
Pawtuxent Toxics (PAWTOXIC), 23, 42
PBPK model, see Physiologically based pharmacokinetic model
PCBs, see Polychlorinated biphenyls
PCDDs, see Polychlorinated dibenzo-*p*-dioxins
PCDFs, see Polychlorinated dibenzofurans
PDM, see Probabilistic Dilution Model
PEC, see Predicted environmental concentration
Photolysis, 37
Physiologically based pharmacokinetic (PBPK) model, 73, 81
Phytoplankton, 66, 71
Pimephales promelas, 75
PLUMES, 39
PNEC, see Predicted no-effect concentration
POC, see Particulate organic carbon
Polyaromatic hydrocarbons (PAHs), 101
Polychlorinated biphenyls (PCBs), 19, 65
    accumulation of in striped bass, 66
    analysis of in Great Lakes food chain, 67
    in Great Lakes food chain, 95
    in Lake Ontario, 69
Polychlorinated dibenzo-*p*-dioxins (PCDDs), 70
Polychlorinated dibenzofurans (PCDFs), 70
Predators, dietary exposure of metal to, 90
Predicted environmental concentration (PEC), 1, 9, 53, 61
Predicted no-effect concentration (PNEC), 10, 54, 61
Probabilistic Dilution Model (PDM), 34
Program Monte, 34
Proton competition, 80

Q

Quantitative Water Air Sediment Interaction (QWASI) model, 20, 21, 35, 103
QWASI model, see Quantitative Water Air Sediment Interaction model

R

RCATOX model, see Row-Column AESOP for Toxics model
RECOVERY, 104

Regulatory agencies, requirements of, 98
RIVEQLII, 42
River Risk (RIVRISK), 17, 35, 42, 104
RIVRISK, see River Risk
Row-Column AESOP for Toxics (RCATOX) model, 29

S

SAB, see Science Advisory Board
Salmo trutta, 69
Salvelinus namaycush, 69
Science Advisory Board (SAB), 78
Sediment(s)
    aerobic bottom, 10
    carbon-normalized, 84
    chemistry model, 99
    diagenesis model, 30
    flux, 6
    modeling of metal toxicity in, 81
    porewater cadmium in, 86
    predictor of metal toxicity in, 85
    toxicity, 22
    transport models, 29, 91, 92
    –water
        column transfers, chemical, 48
        exchange process, 102
        partitioning, 71
SEM, see Simultaneously extracted metal
SEM–AVS
    model, see Simultaneously extracted metal–acid-volatile sulfide model
    ratio, 83
Silver
    BLM-predicted LC50 versus observed LC50 for, 79
    toxicity, 100
Simplified Lake and Stream Analysis (SLSA), 19, 20, 93, 104
Simplified Method Program-Variable-Complexity Stream Toxics (SMPTOX) Model, 25, 104
Simultaneously extracted metal (SEM), 10, 22, 25, 62, 98
Simultaneously extracted metal–acid-volatile sulfide (SEM–AVS) model, 99
SLSA, see Simplified Lake and Stream Analysis
SMPTOX Model, see Simplified Method Program-Variable-Complexity Stream Toxics Model
Soil submodel, 22

Sources of available models, 103–106
    bioaccumulation models, 106
    chemical equilibrium models, 105
    fate and transport models, 103–105
    toxicity models, 106
Spillover hypothesis, metal toxicity, 89
Stand-alone models, 11, 39, 51
Stanford Watershed Model (SWM), 31
Steady-state models, 16, 23, 24, 52
Submodel(s)
    atmosphere, 22
    effects of, 6
    food-chain, 22
    geochemical, 97
    soil, 22
Superfund, 33
Suspended solids, 19
SWM, *see* Stanford Watershed Model
Systems Variability Analysis Code (SYVAC), 22, 35
SYVAC, *see* Systems Variability Analysis Code

T

TBTO, *see* Tributyltin oxide
TDS, *see* Total dissolved solids
Terrestrial nutrient cycling models, 56
TGD, *see* European Commission technical guidance document
Thermodynamic database, 94
Thomann model, 67, 68, 96, 106
Time-variable numerical solution models, 16, 26, 28
TOC, *see* Total organic carbon
Total body burden, Tier I analysis and, 88
Total dissolved solids (TDS), 33
Total organic carbon (TOC), 45
Toxic contaminants, accumulation of by aquatic organisms, 62
Toxicity
    additive, 80
    copper, 76, 100
    models, *see* Bioaccumulation and toxicity models
    silver, 100
TOXIWASP, 29
Tributyltin oxide (TBTO), 23

U

Uniform System for the Evaluation of Substances (USES), 22, 35, 104

U.S. Army Corps of Engineers, 31
U.S. Environmental Protection Agency (USEPA), 96
    -developed WQC, 34
    screening procedures, 52
USEPA, *see* U.S. Environmental Protection Agency
USES, *see* Uniform System for the Evaluation of Substances
U.S. Geological Survey (USGS), 55
USGS, *see* U.S. Geological Survey

V

Volatile species, air–water exchange of, 21
Volatilization, 37

W

WASP, *see* Water Quality Analysis Simulation Program
WASP5, *see* Water Quality Analysis Simulation Program Version 5
Waste load allocations, 51
WASTOX, *see* Water Quality Analysis Simulation of Toxics
Water
    –bed interactions, 42, 49
    column–sediment exchange rates, 26
    –particle exchange process, 102
    quality modeling techniques, 5
Water Quality Analysis Simulation Program (WASP), 27
Water Quality Analysis Simulation Program Version 5 (WASP5), 9, 104
Water Quality Analysis Simulation of Toxics (WASTOX), 29, 44, 93, 104
Water Quality Assessment Methodology (WQAM), 17, 51, 93, 105
Water-quality criteria (WQC), 2, 10, 15, 54, 73
    copper, 47, 96
    statistically based, 61
    USEPA-developed, 34
Watersheds, applicability of HPSF to, 31
WHAM, *see* Windermere Humic Aqueous Model
Windermere Humic Aqueous Model (WHAM), 47, 55, 75, 94, 105
    calibration of to variety of metals, 60
    lack of generic computation framework in, 56
    metal–DOM interactions and, 99
    metal–organic matter interactions in, 56

WQAM, *see* Water Quality Assessment Methodology
WQC, *see* Water-quality criteria

Z

ZID, *see* Zone of initial dilution
Zinc, critical body burden of, 78
Zone of initial dilution (ZID), 41
Zooplankton, 66, 71

SETAC

A Professional Society for Environmental Scientists and Engineers and Related Disciplines Concerned with Environmental Quality

The Society of Environmental Toxicology and Chemistry (SETAC), with offices currently in North America and Europe, is a nonprofit, professional society established to provide a forum for individuals and institutions engaged in the study of environmental problems, management and regulation of natural resources, education, research and development, and manufacturing and distribution.

Specific goals of the society are:

- Promote research, education, and training in the environmental sciences.
- Promote the systematic application of all relevant scientific disciplines to the evaluation of chemical hazards.
- Participate in the scientific interpretation of issues concerned with hazard assessment and risk analysis.
- Support the development of ecologically acceptable practices and principles.
- Provide a forum (meetings and publications) for communication among professionals in government, business, academia, and other segments of society involved in the use, protection, and management of our environment.

These goals are pursued through the conduct of numerous activities, which include:

- Hold annual meetings with study and workshop sessions, platform and poster papers, and achievement and merit awards.
- Sponsor a monthly scientific journal, a newsletter, and special technical publications.
- Provide funds for education and training through the SETAC Scholarship/Fellowship Program.
- Organize and sponsor chapters to provide a forum for the presentation of scientific data and for the interchange and study of information about local concerns.
- Provide advice and counsel to technical and nontechnical persons through a number of standing and ad hoc committees.

SETAC membership currently is composed of more than 5,000 individuals from government, academia, business, and public-interest groups with technical backgrounds in chemistry, toxicology, biology, ecology, atmospheric sciences, health sciences, earth sciences, and engineering.

If you have training in these or related disciplines and are engaged in the study, use, or management of environmental resources, SETAC can fulfill your professional affiliation needs.

All members receive a newsletter highlighting environmental topics and SETAC activities, and reduced fees for the Annual Meeting and SETAC special publications.

All members except Students and Senior Active Members receive monthly issues of *Environmental Toxicology and Chemistry (ET&C)*, a peer-reviewed journal of the Society. Student and Senior Active Members may subscribe to the journal. Members may hold office and, with the Emeritus Members, constitute the voting membership.

If you desire further information, contact the appropriate SETAC Office.

SETAC North America	SETAC Europe
1010 North 12th Avenue	Avenue de la Toison d'Or 67
Pensacola, Florida 32501-3367 USA	B-1060 Brussels, Belgium
T 850 469 1500 F 850 469 9778	T 32 2 772 72 81 F 32 2 770 53 83
E setac@setac.org	E setac@setaceu.org

www.setac.org

Environmental Quality Through Science®

Other titles from the Society of Environmental Toxicology and Chemistry (SETAC):

Code of Life-Cycle Inventory Practice
de Beaufort-Langeveld, Bretz, van Hoof, Hischier, Jean, Tanner, Huijbregts, editors
2003

Contaminated Soils: From Soil–Chemical Interactions to Ecosystem Management
Lanno, editor
2003

Environmental Impacts of Pulp and Paper Waste Streams
Stuthridge, van den Heuvel, Marvin, Slade, Gifford, editors
2003

Life-Cycle Assessment in Building and Construction
Kotaji, Edwards, Shuurmans, editors
2003

Porewater Toxicity Testing: Biological, Chemical, and Ecological Considerations
Carr and Nipper, editors
2003

Reevaluation of the State of the Science for Water-Quality Criteria Development
Reiley, Stubblefield, Adams, Di Toro, Erickson, Hodson, Keating Jr, editors
2003

*Bioavailability of Metals in Terrestrial Ecosystems:
Importance of Partitioning for Bioavailability to Invertebrates, Microbes, and Plants*
Allen, editor
2003

Community-Level Aquatic System Studies—Interpretation Criteria (CLASSIC)
Giddings, Brock, Heger, Heimbach, Maund, Norman, Ratte, Schäfers, Streloke,
editors
2002

Interconnections between Human Health and Ecological Variability
Di Giulio and Benson, editors
2002

Life-Cycle Impact Assessment: Striving towards Best Practice
Udo de Haes, Finnveden, Goedkoop, Hauschild, Hertwich, Hofstetter, Jolliet,
Klöpffer, Krewitt, Lindeijer, Müller-Wenk, Olsen, Pennington, Potting, Steen,
editors
2002

Silver in the Environment: Transport, Fate, and Effects
Andren and Bober, editors
2002

Test Methods to Determine Hazards for Sparingly Soluble Metal Compounds in Soils
Fairbrother, Glazebrook, van Straalen, Tararzona, editors
2002

9th LCA Case Studies Symposium
2001

Avian Effects Assessment: A Framework for Contaminants Studies
Hart, Balluff, Barfknecht, Chapman, Hawkes, Joermann, Leopold, Luttik, editors
2001

Ecological Variability: Separating Natural from Anthropogenic Causes of Ecosystem Impairment
Baird and Burton, editors
2001

Guidance Document on Regulatory Testing and Risk Assessment Procedures for Protection Products with Non-Target Arthropods (ESCORT 2)
Candolfi, Barrett, Campbell, Forster, Grady, Huet, Lewis, Schmuck, Vogt, editors
2001

Impacts of Low-Dose, High-Potency Herbicides on Nontarget and Unintended Plant Species
Ferenc, editor
2001

Risk Management: Ecological Risk-Based Decision-Making
Stahl, Bachman, Barton, Clark, deFur, Ells, Pittinger, Slimak, Wentsel, editors
2001

8th LCA Case Studies Symposium
2000

Development of Methods for Effects-Driven Cumulative Effects Assessment Using Fish Populations:
Moose River Project
Munkittrick, McMaster, Van Der Kraak, Portt, Gibbons, Farwell, Gray, authors
2000

Ecotoxicology of Amphibians and Reptiles
Sparling, Linder, Bishop, editors
2000

Environmental Contaminants and Terrestrial Vertebrates:
Effects on Populations, Communities, and Ecosystems
Albers, Heinz, Ohlendorf, editors
2000

Evaluation of Persistence and Long-Range Transport of Organic Chemicals in the Environment
Klecka, Boethling, Franklin, Grady, Graham, Howard, Kannan, Larson, Mackay, Muir, van de Meent, editors
2000

Multiple Stressors in Ecological Risk and Impact Assessment:
Approaches to Risk Estimation
Ferenc and Foran, editors
2000

Natural Remediation of Environmental Contaminants:
Its Role in Ecological Risk Assessment and Risk Management
Swindoll, Stahl, Ells, editors
2000

7th LCA Case Studies Symposium
1999

Evaluating and Communicating Subsistence Seafood Safety in a Cross-Cultural Context:
Lessons Learned from the Exxon Valdez *Oil Spill*
Field, Fall, Nighswander, Peacock, Varanasi, editors
1999

Ecotoxicology and Risk Assessment for Wetlands
Lewis, Mayer, Powell, Nelson, Klaine, Henry, Dickson, editors
1999

Endocrine Disruption in Invertebrates: Endocrinology, Testing, and Assessment
DeFur, Crane, Ingersoll, Tattersfield, editors
1999

Guidance Document on Higher-Tier Aquatic Risk Assessment for Pesticides (HARAP)
Campbell, Arnold, Brock, Grandy, Heger, Heimbach, Maund, Streloke, editors
1999

Linkage of Effects to Tissue Residues: Development of a Comprehensive Database for Aquatic Organisms Exposed to Inorganic and Organic Chemicals
Jarvinen and Ankley, editors
1999

Multiple Stressors in Ecological Risk and Impact Assessment
Foran and Ferenc, editors
1999

Reproductive and Developmental Effects of Contaminants in Oviparous Vertebrates
Di Giulio and Tillitt, editors
1999

Restoration of Lost Human Uses of the Environment
Grayson Cecil, editor
1999

6th LCA Case Studies Symposium
1998

Advances in Earthworm Ecotoxicology
Sheppard, Bembridge, Holmstrup, Posthuma, editors
1998

Ecological Risk Assessment: A Meeting of Policy and Science
Peyster and Day, editors
1998

Ecological Risk Assessment Decision-Support System: A Conceptual Design
Reinert, Bartell, Biddinger, editors
1998

Ecotoxicological Risk Assessment of the Chlorinated Organic Chemicals
Carey, Cook, Giesy, Hodson, Muir, Owens, Solomon, editors
1998

Principles and Processes for Evaluating Endocrine Disruption in Wildlife
Kendall, Dickerson, Giesy, Suk, editors
1998

Radiotelemetry Applications for Wildlife Toxicology Field Studies
Brewer and Fagerstone, editors
1998

Sustainable Environmental Management
Barnthouse, Biddinger, Cooper, Fava, Gillett, Holland, Yosie, editors
1998

Uncertainty Analysis in Ecological Risk Assessment
Warren-Hicks and Moore, editors
1998

5th LCA Case Studies Symposium
1997

Atmospheric Deposition of Contaminants to the Great Lakes and Coastal Waters
Baker, editor
1997

Biodegradation Kinetics: Generation and Use of Data for Regulatory Decision-Making
Hales, Feijtel, King, Fox, Verstraete, editors
1997

Biotransformation in Environmental Risk Assessment
Sijm, de Bruijn, de Boogt, de Wolf, editors
1997

Chemical Ranking and Scoring: Guidelines for Relative Assessments of Chemicals
Swanson and Socha, editors
1997

Chemically Induced Alterations in Functional Development and Reproduction of Fishes
Rolland, Gilbertson, Peterson, editors
1997

Ecological Risk Assessment for Contaminated Sediments
Ingersoll, Dillon, Biddinger, editors
1997

Life-Cycle Impact Assessment: The State-of-the-Art, 2nd ed.
Barnthouse, Fava, Humphreys, Hunt, Laibson, Moesoen, Owens, Todd, Vigon, Weitz, Young, editors
1997

Public Policy Application of Life-Cycle Assessment
Allen and Consoli, editors
1997

Quantitative Structure-Activity Relationships (QSAR) in Environmental Sciences VII
Chen and Schüürmann, editors
1997

Reassessment of Metals Criteria for Aquatic Life Protection: Priorities for Research and Implementation
Bergman and Dorward-King, editors
1997

Simplifying LCA: Just a Cut?
Christiansen, editor
1997

Workshop of Endocrine Modulators and Wildlife: Assessment and Testing (EMWAT)
Tattersfield, Matthiessen, Campbell, Grandy, Länge, editors
1997

Asking the Right Questions: Ecotoxicology and Statistics
Chapman, Crane, Wiles, Noppert, McIndoe, editors
1996

Pesticides, Soil Microbiology and Soil Quality
Anderson, Arnold, Malkomes, Lagacherie, Oliveira, Plicken, Tarry, Soulas, Torstensson, editors
1996

Towards a Methodology for Life-Cycle Impact Assessment
Udo de Haes, editor
1996

*Whole Effluent Toxicity Testing:
An Evaluation of Methods and Prediction of Receiving System Impacts*
Grothe, Dickson, Reed-Judkins, editors
1996

Guidance Document on Regulatory Testing Procedures for Pesticides with Non-Target Arthropods
Barrett, Grady, Harrison, Hassan, Oomen, editors
1995